T0265425

CAMBRIDGE LIBRARY COLLECTION

Books of enduring scholarly value

Technology

The focus of this series is engineering, broadly construed. It covers technological innovation from a range of periods and cultures, but centres on the technological achievements of the industrial era in the West, particularly in the nineteenth century, as understood by their contemporaries. Infrastructure is one major focus, covering the building of railways and canals, bridges and tunnels, land drainage, the laying of submarine cables, and the construction of docks and lighthouses. Other key topics include developments in industrial and manufacturing fields such as mining technology, the production of iron and steel, the use of steam power, and chemical processes such as photography and textile dyes.

A Short History of the Steam Engine

A Short History of the Steam Engine, first published in 1939, remains one of the most readable and clear explanations of the topic for the non-specialist. H.W. Dickinson limits himself to stationary engines and boilers, and only touches on the beginnings of locomotive and marine engines. He puts the stages of development in their context, showing how economic and social factors were involved in the evolution of the steam engine. The illustrations are plentiful and the text, while technical, never becomes impenetrable. The successive improvements to the simple engines of the seventeenth century, as new materials or purposes arose, are developed chapter by chapter to the twentieth century. Each engineer was building on the work of his predecessors, rather than there being any single inventor of genius. Dickinson also wrote biographies of key figures of the Industrial Revolution, which are being reissued in this series.

Cambridge University Press has long been a pioneer in the reissuing of out-of-print titles from its own backlist, producing digital reprints of books that are still sought after by scholars and students but could not be reprinted economically using traditional technology. The Cambridge Library Collection extends this activity to a wider range of books which are still of importance to researchers and professionals, either for the source material they contain, or as landmarks in the history of their academic discipline.

Drawing from the world-renowned collections in the Cambridge University Library, and guided by the advice of experts in each subject area, Cambridge University Press is using state-of-the-art scanning machines in its own Printing House to capture the content of each book selected for inclusion. The files are processed to give a consistently clear, crisp image, and the books finished to the high quality standard for which the Press is recognised around the world. The latest print-on-demand technology ensures that the books will remain available indefinitely, and that orders for single or multiple copies can quickly be supplied.

The Cambridge Library Collection will bring back to life books of enduring scholarly value (including out-of-copyright works originally issued by other publishers) across a wide range of disciplines in the humanities and social sciences and in science and technology.

A Short History
of the Steam Engine

HENRY WINRAM DICKINSON

CAMBRIDGE
UNIVERSITY PRESS

CAMBRIDGE UNIVERSITY PRESS

Cambridge, New York, Melbourne, Madrid, Cape Town, Singapore,
São Paolo, Delhi, Dubai, Tokyo, Mexico City

Published in the United States of America by Cambridge University Press, New York

www.cambridge.org
Information on this title: www.cambridge.org/9781108012287

© in this compilation Cambridge University Press 2010

This edition first published 1939
This digitally printed version 2010

ISBN 978-1-108-01228-7 Paperback

Additional resources for this publication at www.cambridge.org/9781108012287

A SHORT HISTORY
OF THE
STEAM ENGINE

CAMBRIDGE
UNIVERSITY PRESS
LONDON: BENTLEY HOUSE
NEW YORK, TORONTO, BOMBAY
CALCUTTA, MADRAS: MACMILLAN
TOKYO: MARUZEN COMPANY LTD

DENIS PAPIN
1647–1712?

JOHN SMEATON
1724–92

JAMES WATT
1736–1819

OLIVER EVANS
1755–1819

RICHARD TREVITHICK
1771–1833

PIONEERS OF THE RECIPROCATING ENGINE

A SHORT HISTORY
OF THE
STEAM ENGINE

by

H. W. DICKINSON

Author of
'Robert Fulton, Engineer and Artist',
'James Watt, Craftsman and Engineer',
'Matthew Boulton', etc.

CAMBRIDGE
AT THE UNIVERSITY PRESS
1939

PRINTED IN GREAT BRITAIN

CONTENTS

PART I. THE RECIPROCATING ENGINE

Man's apprenticeship of toil—Minerals pressed into his service—
Dark Ages—Renaissance of learning—Establishment of Scientific
Societies—Where was power to be sought?—Discovery that the
atmosphere has weight—Von Guericke's air pump—Scientific specula-
tion—Papin makes a vacuum under a piston by condensing steam—
Practical but ineffective efforts of De Caus, Ramsay, Marquis of
Worcester, and Morland—Recapitulation.

Savery employs pressure and suction for raising water—His patent
and its extension—Great promise but failure for mining—Used for
domestic supply—Papin makes a Savery engine—Blakey makes an
automatic one.

Newcomen's engine as early as Savery's but fundamentally different—
Mårten Triewald's story—Early difficulties surmounted—Detractors
—Dudley Castle engine—Death of Savery—His patent acquired by
certain "Proprietors"—Newcomen checked—Expiration of Savery's
patent—Spread of the engine in Great Britain and on the Continent
—Death of Newcomen.

Rebirth of iron smelting in England—Increase in size of engines—
Boiler capacity—Cost of erection and operation—Smeaton's experiments—His improved engine—His water coal-gin—The atmospheric
engine modified for winding from mines.

Watt's career—Repairs atmospheric engine model—Stumbles on
latent heat—Invents and patents separate condenser—Roebuck helps

with experimental engine—Boulton takes over patent—Watt removes
to Birmingham—Extension of patent—Partnership with Boulton—
Engines in Cornwall—Premium—Compound engine—Rotative
engine—Substitutes for crank—Governor—Expansive working—
Indicator—Infringements—Partnership expires—Watt summed up
—Economic advance.

PART II. THE STEAM TURBINE

LIST OF PLATES

LIST OF FIGURES IN THE TEXT

*This plate is available for download from
www.cambridge.org/9781108012287

TAILPIECE

TABLES

PREFACE

MANY even well-informed persons believe that the steam engine is out of date. They say that electricity has taken its place, because, nowadays, wherever they go, in factory, workshop and even on railways, a steam engine is hardly ever to be seen; instead they see electric motors driving the necessary machinery, and if not that type of motor, then they see gas, oil and petrol engines in use. Superficially this is true, but the hard fact is far otherwise—the steam engine has not been displaced, as engineers know full well; they know that the current that enables these electric motors, seen and unseen, to function requires for its generation huge engines in large power stations, and that these engines, known as turbo-generators, are actuated very largely by steam derived from coal.

To support this statement we take leave to introduce a few statistics, dry but useful. World power production,[1] that is for manufacturing, lighting, heating, railways, motor cars, ships, etc., reduced to a common basis in the same unit of energy and differentiated according to the sources whence the power is derived, is estimated as follows:

	Millions of kilowatt-hours	Percentage
Hard coal	1,276,400	77·0
Brown coal	34,600	2·1
Oil	260,000	15·7
Water power	84,800	5·1
Total	1,655,800	99·9

These figures are astronomical; one way to realize their import is to say that power is being generated every hour of the day and every day of the year at an average rate of one-tenth of a kilowatt for each man, woman and child in the whole world.

Another angle of approach to the question is to look at the annual statistics of production of electrical supply undertakings,[2]

[1] *Power Resources of the World*, World Power Conference, 1929, p. 82.
[2] *Statistical Year Book* of the World's Power Conference, No. 1 (1933–4), 1936, Abstract of pp. 96, 97.

a much more restricted field, of course. Adding together the installed capacity of plants in Europe, North, Central and South America, Africa, Asia, and Australasia, expressed in thousands of kilowatts, we have:

		Percentage
Fuel operated plants	54,907	62·8
Internal combustion operated plants	1,543·5	1·8
Hydro-electric river flow plants	27,988	32·0
Hydro-electric river storage plants	3,001	3·4
	87,439·5	100·0

It is a notable fact that the United States' share of the total is 39 per cent. These figures are higher to-day.

The truth is that the steam engine was never so important in the world's economy as it is to-day. It is not, therefore, as it might appear, merely an academic exercise to write about its history, and no excuse is offered for treating the theme once again. It is appropriate that it should be done in this country where so many major steam-engine inventions have been matured.

This volume does not pretend to be more than its title indicates, viz. a short history; no one is more conscious of its shortcomings than the author. In view of this limitation the treatment has been restricted to the history of the stationary engine, not neglecting that of the steam boiler, which is really part and parcel of the whole. The attempt has been made to compress within narrow limits the essential facts of the history and to present these with due balance, and in relation to some of the material changes, economic and social, that have accompanied or followed the development of the engine. In addition to what appears in the text, essential facts will be found tabulated in a Synopsis, facing p. 248 (Pl. X). The marine engine and the locomotive have been touched upon only up to the point where they branch off from the parent stem and begin their independent careers. To stop short there seemed to the author to be justifiable as the marine engine has been exhaustively dealt with in a recent work[1] and the locomotive has never lacked historians owing to the fascination it exerts, especially over the layman; several recent authoritative works on its history are in existence.

[1] Smith, E. C., *History of Marine Engineering*, 1938.

The author hopes that the present volume will appeal, in the first place, to the many engineers who desire to know something of the history of an important branch of their profession, and one that impinges on nearly every other branch. In the second place he hopes to interest the general public, especially the younger generation, and in doing so he wishes he may be able to clear away some of the errors into which previous writers have fallen, and which have been repeated time and again; in particular the author will feel that his labours have not been in vain if he succeeds in uprooting the fallacies that James Watt invented the steam engine, and George Stephenson the locomotive.

The author has to confess that the writing of a history of the steam engine has been a dream of his of more than thirty years standing, ever since his preoccupation with the subject at the Science Museum, South Kensington, London, but it is doubtful whether he would have undertaken the task had it not been for the encouragement of his old friend, Mr Rhys Jenkins. It is to the researches of the latter into the early days of the heat engine, in the seventeenth and eighteenth centuries, that we owe the light that has been thrown on the darkness that had hitherto largely enveloped that period.

Once the book was started, the author received help from many quarters, particularly from fellow Members of the Newcomen Society for the Study of Engineering and Technology, to the Council of which he is indebted for permission to make unrestricted use of its *Transactions*. Quondam colleagues at the Science Museum, among whom the author must mention Mr A. Stowers, could not have done more to forward the work had it been a Museum publication, which in a sense it is. Mr A. A. Gomme of the Patent Office and his courteous Library Staff have given invaluable help. The Royal Society has permitted the use of original material. The Institution of Mechanical Engineers has allowed the use of much matter and illustrations from its *Proceedings*. It is almost invidious to particularize those friends of the author who have assisted him with material, but he cannot omit to mention Engineer Captain E. C. Smith, R.N. ret., whose store of biographical and technical information is perhaps unrivalled. Dr Gerald B. Stoney has permitted lengthy quotations

from his Parsons Memorial Lecture; Mr R. H. Parsons has been
of the greatest assistance in the history of the steam turbine of
which he has a deep knowledge; Professor E. N. da C. Andrade
has been most stimulating. Mr Carl W. Mitman of the National
Museum of Washington and Mr Greville Bathe have supplied
information from the United States. Mr R. E. Trevithick, a
descendant of the great engineer, has also supplied information.
Dr Ing. Birger Ljungström has furnished invaluable details about
his turbine and himself. Criticisms of the author's son, H. D.
Dickinson, M.A., at the MS. stage were penetrating and con-
structive. The Editors of *The Engineer* and of *Engineering* have
been generous in allowing the use of material from their pages.
Several gentlemen who have helped greatly, and whom the
author would have liked to have thanked by name, desire to
remain anonymous.

Many firms have been appealed to for information and have
responded readily. The author acknowledges the help of the
following: Aktiebolaget de Lavals Ångturbin, Stockholm;
Babcock & Wilcox Ltd., London and Renfrew; Belliss & Morcom
Ltd., Birmingham; British Brown-Boveri Ltd., London; Peter
Brotherhood Ltd., Peterborough; Brush Electrical Engineering
Co. Ltd., Loughborough; General Electric Co., Schenectady;
Greenwood & Batley Ltd., Leeds; La Mont Steam Generator
Ltd., London; C. A. Parsons & Co. Ltd., Newcastle-upon-
Tyne; La Société Rateau, Paris; Robey & Co. Ltd., Lincoln;
Westinghouse Electric & Manufacturing Co., Pittsburgh;
United States Metallic Packing Co. Ltd., Bradford.

Messrs E. & F. N. Spon Ltd. and Messrs Longman Green
& Co. have permitted the use of book illustrations which will
be found acknowledged.

The staff of the Cambridge Press have helped greatly in the
preparation of the illustrations and have striven with the author,
as aforetime, for perfection in production.

H. W. D.

Purley, Surrey

PART I

THE RECIPROCATING ENGINE

CHAPTER I

INTRODUCTORY

Man's apprenticeship of toil—Minerals pressed into his service—Dark Ages —Renaissance of learning—Establishment of Scientific Societies—Where was power to be sought?—Discovery that the atmosphere has weight— Von Guericke's air pump—Scientific speculation—Papin makes a vacuum under a piston by condensing steam—Practical but ineffective efforts of De Caus, Ramsay, Marquis of Worcester, and Morland—Recapitulation.

THE heat engine, undoubtedly the most powerful agent in bringing about the material well-being that everyone shares, or ought to share, to-day, is comparatively a recent comer into the world, being in fact, little more than two and a quarter centuries old. This is such a short time as compared with the eight to twelve thousand years of human civilization that curiosity is aroused, so sophisticated are we to-day, to find an explanation why it was the heat engine did not arrive on the scene earlier. To do so we must cast our thoughts back and try to recapture the outlook of mankind before its introduction, to examine what made the want of it felt, and what were the experiments and reasonings that finally brought the heat engine into being. Let it be said then that during the long period alluded to, a period justly termed by one writer "Man's apprenticeship of toil",[1] Man was winning control slowly, but at an ever-increasing rate over the powers of nature. He had learnt to make fire, to shape implements and tools; he had acquired a fund of transmitted empirical knowledge; he had schooled himself in craftsmanship; he had discovered and applied the mechanical powers: the lever, the inclined plane, the wheel and axle, the pulley; he had extended his own bodily powers by domesticating and harnessing animals for draught and other purposes; he had learnt to take advantage of the wind to drive ships by sails, and (much later) to grind his cereals by windmills. Long after he had learnt how to make fire, he began to acquire

---

[1] Vowles, *The Quest for Power*, 1931, p. 1.

BSH

I

knowledge of minerals and ores, to make progress in reducing them to the metallic state, and to forge or cast the resultant metals into useful objects. Thus was initiated a subtle change of the most far-reaching consequences. Hitherto man had been an animal like others, finding all he needed for his sustenance reproducible on the earth's surface. Now he extended the domain required for his support to products below the surface. The former is an economy that can subsist, it will be realized, as long as the earth remains habitable; the latter economy, in which we exist to-day, is bound to be short in duration as compared with the former, because we are using up non-reproducible assets— an arresting thought.

Another large factor in material progress that has to be taken into account is division of labour, widening from the family to the tribe, then to other tribes, bringing about the production of commodities by specialization, leading to barter and then to exchange for money, hence giving rise to trade and commerce overland and overseas. The Asiatic trade routes and the Mediterranean afford examples of growth of this kind of commerce for at least two centuries B.C. onwards, to the growth of cities like Bagdad and Alexandria, to the cultivation of learning in schools, and to the spread of technology by the immigration of artisans. Now at last one would imagine that the first glimmerings of steam power should show themselves. This was the case, as the writings of Heron of Alexandria show; but the applications of steam were to produce what we might almost term puerile effects to astonish or awe the ignorant rather than to do useful work.[1] The fact was that the conditions and abundance of slave labour for performing menial tasks obviated any insistent demand for other than animal power, even by the Romans with their far-flung organization.

The apparent stagnation of the Dark Ages, and the slow recovery of learning during the Middle Ages, interfused and fecundated by the learning of the Arabians based upon Greek thought, brought about the greatest outburst of the human spirit that the world has ever known. We have only to reflect

[1] See p. 185.

upon the influence of the mariners' compass, the discovery of a New World, the invention of printing by movable type, and the spread of the art of paper-making in the Western World, to realize what large factors these were in helping to bring about the result. It was a world with a complete change in the attitude of mind and of ideals of life, covering every field of knowledge, art, politics and religion. In the economic sphere the decay of the feudal system had been accompanied by the rise of a commercial class, and this led to the mercantilist policy steadily pursued by sovereigns such as the Tudors and directed towards the attainment of military power. This led to the introduction and fostering by legislative action of new industries, and to the immigration of artisans to carry them on.

In no domain was the outcome of the Renaissance more fruitful than in that of science and technology. The spirit of enquiry was aroused; observation and experiment instead of speculation and dogmatism became the accepted approach to any problem. Throughout Europe, societies and academies for the study of science sprang up by the hundred, as for example our own Royal Society in 1662. This ferment of ideas led to new wants and to attempts to satisfy them. It is significant that the Act consolidating the law laying down the principles for granting of Letters Patent for inventions in this country—the Statute of Monopolies—was passed in 1624.

The change that we have remarked upon, from a purely agricultural to an accessory mineral economy, had not proceeded far before the exhaustion of surface ore deposits led to search for the source of such ores in the mineral veins. Then began outcrop or adit working, and the sinking of pits to obtain access to the mineral seams. This form of mining did not progress far before that bugbear of the miner—water—was encountered. For the higher seams, adit drainage was possible, but for situations below natural drainage level, existing appliances—hand pumps, animal-driven chains of buckets and such like—had to be employed, but they were woefully inadequate besides being inefficient. Plainly some more powerful agent than that of animals was needed. The problem was to draw water from a considerable depth, or to force it to a moderate height, or perhaps to do both.

But in which of the hundred and one possible directions could help be sought? To the Greek mind there were four elements: earth, air, fire and water, and one or more of these must be enlisted; but where to begin? Earth connoted the material that was to be won from the ground, water certainly could be utilized to raise other water by means of a waterwheel, but this was clumsy and only rarely were the necessary conditions to be found together at any one place. There remained air and fire. The former and its effects when in motion were matters of everyday observation; the wind, as we have said, had been harnessed to do work by means of the windmill as early as the twelfth century, but its power can be exerted only in high places generally far removed from mines and is intermittent. The use of fire to heat water in a vessel, giving rise to ebullition and the disappearance of the water in the form of visible vapour, gives the impression of considerable energy, but it does not suggest a means of controlling it, still less of raising water by its means. It was known also that if water were heated in a confined space a pressure would be reached that could force out a jet of steam or water through a tube or else, if there were no outlet, become destructive in its effects. The pressure of steam must therefore have been deemed a likely avenue to explore. We may cite, for example, the laboratory experiment, for it seems to have been nothing more, of Giambattista della Porta (1538–1615), natural philosopher of Naples. In his apparatus, described and illustrated in his *Spiritali*, 1606 (see Fig. 1), he employs a fireplace *F*, generating steam, in what looks like a retort, or it may be a wine flask *D*, communicating with a closed chamber *B* above the surface of water therein contained; the steam forces out the water through the tube *C*. The size of the apparatus may be inferred from Porta's statement that the retort held two ounces of water. We have singled out Porta because he shows another laboratory experiment. A retort or wine flask *A*, like the one in the preceding experiment, when full of steam is plunged under water contained in a bowl, a vacuum is formed and water is drawn up into the flask. Porta is the first writer, so far as is known to the author, to show that water could be raised in this way. There is no evidence that these experiments had any practical outcome; all we can say is that they showed great originality.

The pressure of steam was not, however, the only avenue to explore. Gunpowder was already applied in fire-arms, and its explosive effect was an obvious source of energy, could some way be found of moderating its violence and repeating the action regularly: even so the way to apply the resulting pressure to raise water was only a speculation.

Then there was the air of the atmosphere; here, had it been understood, was the key to the problem of making a steam engine. The great discovery about the atmosphere, namely, that it had weight, is said to have been the outcome of attempts in 1641 by the engineers of Cosmo de' Medici II, Grand Duke of Tuscany, to make a special sucking pump to draw water from a depth, under the bucket, of about 50 ft. They found of

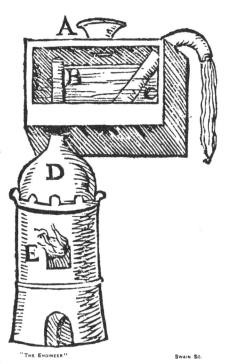

Fig. 1. Della Porta's steam pressure apparatus.

From his *Spiritali*, 1606.

course that try as they would, they could not get water to follow the bucket to a greater height than about 28 ft. from the surface of the water, as in fact they knew from previous experience. The accepted view of the time was that nature abhorred a vacuum; but, if that were the case, there seemed no reason why there should be a limit to the height the water would follow the bucket. Galilei (1564–1642), appealed to for an explanation, suggested lamely that nature's abhorrence of a vacuum ceased when the bucket reached a height of 28 ft. above the water! This dictum may have satisfied the engineers, but we doubt it; it certainly did not satisfy the philosopher himself, and he was led to carry out some experiments which, however, did not resolve his doubts. He confided his speculations to his amanuensis and pupil Evangelista Torricelli (1608–47) who repeated the pump experiment, without, however, getting the water to rise any higher than before. Hence he deduced that the atmosphere had weight and that 28 ft. was the height of a column of water that the weight of the atmosphere balanced, and hence that the latter would balance a proportionately shorter column of a heavier liquid. He tried this with mercury in a closed tube, inverted into a basin of the same liquid, and found his surmise was correct—the mercury would not stand higher than about 28 in. As the specific gravity of mercury is about fourteen times that of water, the theory was strikingly proved; incidentally this was the invention of the mercurial barometer. Torricelli published his result in 1643, and his surprising experiments were everywhere repeated, among others by Blaise Pascal (1623–62), French divine and savant. He carried the experiment a stage further; he reasoned that, the higher one ascended in the atmosphere, the lower should be its pressure. In 1647 he induced his brother-in-law, Florin Périer, to take a mercurial barometer to the top of the Puy de Dôme in Auvergne, a height of 4800 ft. (1465 m.). The result was as anticipated—as Périer ascended, the mercury fell steadily and at the top was roughly 3 in. lower than at the bottom. The fact was thus established that the pressure of the atmosphere varied inversely as the height. At sea-level the pressure of the atmosphere was equal to between 14 and 15 lb. on the square inch.

Another experimenter with the pressure of the atmosphere was Otto von Guericke (1602–86), burgomaster of Magdeburg. About 1650 he applied existing knowledge about pumps to make one for drawing air instead of water, and with it got some spectacular results. He emptied a copper ball of air, which consequently crumpled up "as a cloth is crushed between the fingers". In 1654 at the Diet of Ratisbon, he exhibited some striking experiments to show how great was the pressure of the atmosphere.[1] He had a cylinder some 15 in. diameter by 20 in. high, with an accurately fitted piston (see Fig. 2) to which by a rope over a pulley the joint efforts of fifty or more strong men could be applied. They were unable to pull the piston more than halfway up the cylinder, and when a glass receiver made vacuous by his air pump was applied at X, and the stopcock opened, the piston was forced down and the men were overcome. Another experiment was to lift weights. The same apparatus was used (Fig. 3). The piston being at the top of the stroke was attached by a rope over pulleys to a scale pan carrying 2868 lb. A small boy, twelve to fifteen years of age, was able, by a syringe applied to the stopcock X, to exhaust the air and raise the weight off the ground.

Thus an avenue of approach to making an engine, if a vacuum could be produced readily, loomed in sight.

As a result of von Guericke's experiments, the air pump became a favourite, nay almost a fashionable adjunct to research. The Hon. Robert Boyle (1627–91), seventh son of Richard, Earl of Cork and Orrery, had, as was then usual with sons of gentlemen, travelled on the Continent (1639–44); he had heard of Torricelli's experiments, and also knew what von Guericke had done. Boyle, with the help of Robert Hooke (1635–1703), the well-known physicist and mathematician, succeeded in making an improved air pump. It was finished in 1659, and with it Boyle carried out a series of experiments "touching the weight and spring of the air and its effects", an account of which he published in 1660. In 1667 he, along with Hooke, still further improved the air pump, and with this new one carried out some more experiments of which he published an account in 1669. It is this pump that is still preserved by the Royal Society at Burlington House.

[1] Ottonis de Guericke, *Experimenta Nova . . . Magdeburgica de Vacuo Spatio,* 1672. The title of Fig. 2 is "De vaso vitreo, plures quam viginti vel 50 & plures homines robustos attrahere valente." That of Fig. 3 is "Experimentum ingens Pondus elevandi".

We have mentioned already the establishment throughout Europe of learned societies; their influence on the progress of science soon became manifest. The French Académie Royale des Sciences, established by Louis XIV in Paris in 1666, at the instigation of his minister, Jean Baptiste Colbert (1619–83), was fortunate at the outset in securing the services of Christiaan

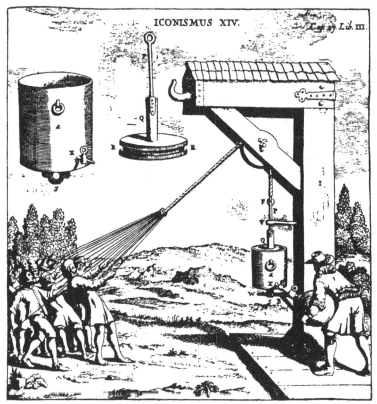

Fig. 2. Von Guericke's air pump overcoming fifty men, 1654.
From his *Experimenta Nova...Magdeburgica*, 1672.

Huygens (1629–95), the celebrated Dutch astronomer, as one of the sixteen Founder Members of the Academy, and for fifteen years he worked for it in Paris. It was on his recommendation that Denis Papin (1647–1712?), doctor of medicine, was appointed in 1671 assistant in the laboratory of the Academy. Huygens started him on a programme of experiments on the weight of air with the air pump, and also on the estimation of the force of gunpowder. Papin, in 1674, published a memoir, "Nouvelles expériences du Vuide, avec la description des machines qui servent à les faire", without, however, having made much advance in the direction of applying the vacuum to practical purposes. In 1675, Papin, influenced by the hope of getting a better position, and possibly by the growing menace

Fig. 3. Von Guericke's air pump applied to raise weights, 1654.
From his *Experimenta Nova Magdeburgica*, 1672.

of religious persecution, for he was a Huguenot, left Paris for London, where he was successful in obtaining employment under Boyle in making experiments with a double-acting air pump of Papin's own construction. The new features of this pump were: valves opened and closed by the action of the machine itself, pistons operated by the feet in stirrups, and pistons covered with water to render them air-tight. These experiments lasted from

July 1676 until February 1679. In the following May, Hooke introduced Papin to the Royal Society, where he showed his apparatus for cooking under pressure in a closed vessel, the well-known "new Digester or Engine for softening Bones", which engaged his attention at intervals for several years. It is not our purpose, however, to follow Papin's career in London, Venice, Paris, Marburg and Cassel, although during this period he produced many ingenious devices based on vacua produced by an air pump or by the explosion of gunpowder, and applied these devices to lifting water for fountains, raising weights from a depth, and to the transmission of power—purposes indicating the urge felt by this remarkable man to produce something practical to meet the needs that had already made themselves felt in the industrial field.[1] We must content ourselves with a notice of his apparatus in which the production of a vacuum by the condensation of steam under a piston was suggested for the first time. This memoir is entitled: "Dion. Papini Nova methodus ad vires motrices validissimas levi pretio comparandas" (Papin's new method of obtaining very great moving powers at small cost).[2]

Papin had been working, as we have said, on the gunpowder engine, but to no purpose, because he found "always after the flame of the gunpowder is extinguished about a fifth part of the air remains in the tube"; he endeavoured to get over this difficulty by using a working agent that did not leave behind it any residue,

since it is a property of water that a small quantity of it turned into vapour by heat has an elastic force like that of air, but upon cold supervening is again resolved into water, so that no trace of the said elastic force remains, I concluded readily that machines could be constructed wherein water, by the help of no very intense heat, and at

[1] For a digested account of Papin's work bearing on the steam engine see Galloway, R. L., *The Steam Engine and its Inventors*, 1881, pp. 14–51.

[2] *Acta Eruditorum Lipsiae*, 1690, pp. 410–14 (Latin was then the universal language of the learned). The Latin text is reprinted in Muirhead, *Mechanical Inventions*, III, p. 139, with an English translation. The French text, the language probably in which it was first written, appeared in *Recueil des diverses pièces*, 1695.

little cost, could produce that perfect vacuum which could by no means be obtained by the aid of gunpowder....Among various constructions which might be contrived for this purpose the following seemed to me the most suitable.

Papin goes on to describe the apparatus (see Fig. 4) which consisted of a tube A about $2\frac{1}{2}$ in. diameter, fitted with a piston BB; in the piston rod was a notch H, into which a catch EE sprung when the piston had reached the top of its stroke. Water to the depth of about $\frac{1}{3}$ in. was placed in the bottom of the tube, fire was applied to the under side, and the water was turned into steam, which thus forced up the piston. When cool, the piston was unlatched and the pressure of the atmosphere forced it down to the bottom of the tube. Papin found "that one minute's time is sufficient for a moderate fire to drive the piston...up to the top", but he does not say how long it took to cool and hence how many strokes were possible, say per hour. The principal difficulty that he foresaw was "in finding a workshop capable of easily making very large tubes". Obviously, we have here little more than a lecture-table apparatus; all the means for turning it into a practical working engine are absent.

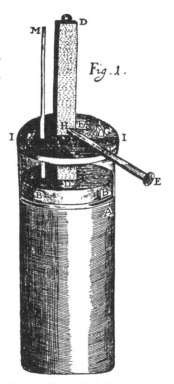

Fig. 4. Papin's cylinder and piston vacuum apparatus.

From *Acta Eruditorum*, 1690.

Here we must leave Papin, with regret that he did not follow up this line of investigation. His subsequent history is a sad one for he was reduced to poverty and died we do not know when and where, possibly in London. Our portrait (see frontispiece) is taken from an oil painting in the University of Marburg, painted probably in 1695 when he was 48 years of age.

Let us now turn back a little way to see what practical men

were doing to solve the ever-arising problem of raising water,
the outcome of which was the invention of the steam engine.[1]
We have no need to go farther back than the seventeenth century,
because it was not until then that the first glimmerings of light
appear, when, as we have stated, a race of engineers concerning
themselves with the supply and use of water was growing up.

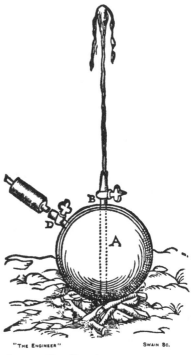

The first person who ought to be
mentioned is Salomon de Caus,[2]
a French landscape gardener,
much concerned with water effects—cascades, fountains and
canals. In the course of his employment he was engaged successively at the Courts of Italy,
England and Germany. In 1611
he was at Greenwich Palace and
at Richmond, where he came
under the notice of Prince Henry,
eldest son of King James I. In
1615 he was at Heidelberg, and
published at Frankfurt an encyclo-
paedic work upon his profession;
out of the whole work he devoted
a hundred words and a single
woodcut (see Fig. 5) to what
he described as "Le troisième
moyen de faire monter [i.e. l'eau]
est par l'aide du feu, dont il se
peut faire diverses machines: j'en

Fig. 5. De Caus's steam pressure
ball.

From his *Les Raisons des Forces
Mouvantes*, 1615.

donneray ici la demonstration d'une". The apparatus is simply
a ball or sphere with a pipe from the inside reaching almost to
the bottom and projecting outside; when partly filled with water

---

[1] For much of this matter the writer is indebted for inspiration to a paper
by Rhys Jenkins, "The heat engine idea in the seventeenth century", *Trans.
Newcomen Soc.* XVII, p. 1.

[2] For details of his career see Maks, Dr C. S., *Salomon de Caus* (1576–
1626), 1935.

and put on a fire, the steam given off forces the remaining water out through the pipe—in other words, it was a steam-pressure apparatus. There is no hint that it was novel; it is not spoken of as being of any importance; in fact, it was only one of a number of pretty "effects". We surmise, however, that he had with this toy sown good seed. In the English Court at that time there was a young Scotsman, named David Ramsay, who could hardly help having known De Caus. Ramsay was a prolific patentee, and in four of his patents means for raising water are included. His patent of 1631 (No. 50) covers a number of inventions, but only three of them call for mention:

To Raise Water from Lowe Pitts by Fire.

To Make any Sort of Mills to goe on Standing Waters by Continual Moc'on without the Helpe of Windes, Waite or Horse.

To make Boates, Shippes and Barges to goe against stronge Winde and Tyde.

We hear nothing further of Ramsay's projects; obviously he had essayed tasks beyond the state of knowledge and of the mechanical arts of the day.

Attending the Court at the time Ramsay obtained these grants, there was a young man, Edward Somerset (1601–67), who was later to become the second Marquis of Worcester, and it can hardly be doubted that he would become acquainted with Ramsay, and learn what the latter was trying to do. Whether it was this acquaintance or simply that the subject of water raising was a much-discussed one at the time, the Marquis's attention was drawn to the subject and he certainly achieved something. It is a question, however, whether he would have been heard of if he had not written a book[1] in which, among other inventions, he describes his steam apparatus (Invention No. 68) but in an obscure manner—intentionally without doubt—that has led to endless conjecture concerning what he actually achieved.

The wording of his description of his apparatus is as follows:

An admirable and most forcible way to drive up water by fire, not by drawing or sucking it upwards, for that must be as the Philosopher calleth it, *Intra sphaeram activitatis*, which is but at such a distance.

[1] *Century of... Inventions*, 1663.

But this way hath no Bounder, if the Vessels be strong enough; for I have taken a piece of a whole Cannon, whereof the end was burst, and filled it three quarters full of water, stopping and scruing up the broken end; as also the Touch-hole; and making a constant fire under it, within 24. hours it burst and made a great crack: So that having a way to make my Vessels, so that they are strengthened by the force within them, and the one to fill after the other. I have seen the water run like a constant Fountaine-stream forty foot high; one Vessel of water rarified by fire driveth up forty of cold water. And a man that tends the work is but to turn two Cocks, that one Vessel of Water being consumed, another begins to force and refill with cold water, and so successively, the fire being tended and kept constant, which the self-same Person may likewise abundantly perform in the interim between the necessity of turning the said Cocks.

Later, in the *Century*, the Marquis describes another engine (Invention No. 98) which has been thought by many to be a steam engine. Although the description is more involved and obscure than the preceding, there is no mention of either fire or steam and in the author's opinion it refers to some form of mechanical transmission. The description reads as follows:

An Engine so contrived, that working the *Primum mobile* forward or backward, upward or downward, circularly or cornerwise, to and fro, streight, upright or downright, yet the pretended Operation Continueth, and advanceth none of the motions above mentioned, hindering, much less stopping the other; but unanimously, and with harmony agreeing they all augment and contribute strength unto the intended work and operation And therefore I call this A *Semi-omnipotent Engine* and do intend that a Model thereof be buried with me.

We mentioned above that the Marquis did achieve something. Here is the evidence.

Samuel Sorbière, historian to the King of France while on his travels in England, says:[1] "One of the most curious things I wished to see was a Hydraulic Machine which the Marquis of Worcester has invented and of which he has made an experiment." Sorbière accordingly visited Vauxhall to view the machine there, and his remark is: "This machine will raise to

[1] *Relation d'un voyage en Angleterre*, Paris, 1664. Translated as *A Voyage to England...*, 1709.

the height of 40 feet by the strength of one man and in the space
of one minute of time, four large buckets of water, and that by
a pipe or tube of 8 inches.''

In 1669, Cosmo de' Medici III, Grand Duke of Tuscany,
came to London and saw the machine. The description of his
visit is as follows:[1]

His highness, that he might not lose the day uselessly, went again
after dinner to the other side of the city, extending his excursion as
far as Vauxhall, beyond the palace of the Archbishop of Canterbury,
to see an hydraulic machine invented by My Lord Somerset, Marquis
of Worcester. It raises water more than forty geometrical feet by the
power of one man only; and in a very short space of time will draw
up four vessels of water through a tube or channel not more than a
span in width; on which account it is considered to be of greater
service to the public than the other machine at Somerset House.

If steam had been used, it is curious it is not mentioned.
The consonance of language between the description of Invention 68 and these two travellers' tales leaves us in little doubt
that one and the same apparatus is concerned in all three. Mr
W. H. Thorpe, a recent writer,[2] went into the question and
assuming a reasonable capacity for the bucket concluded: ''the
mechanical work done was equal to 4800 ft. lb. per minute, an
effort well within the capacity of one man for a short period,
allowing a reasonable efficiency for the apparatus.''

One further point: the Marquis invented also a ''water com-
manding engine'', but whether this was Invention 68, 98 or
perhaps something else entirely, is not known. He secured the
monopoly of his invention or idea by Act of Parliament 15 Car.
II, cap. xii, June 3, 1663,[3] entitled: ''*An Act* to enable *Edward,
Marquess of Worcester* to Receive the Benefit and Profit of a
Water-Commanding Engine by him invented...'' for a term
of 99 years. The Act does not say that the force of fire is to be

---

[1] *Travels of Cosmo the Third, Grand Duke of Tuscany, through England
during the reign of King Charles the Second, translated from the Italian Manu-
script in the Laurentian Library at Florence*, 1821, p. 325.

[2] *Trans. Newcomen Soc.* XIII, p. 76.

[3] For the text of the Act, cf. Dircks, *Life of the Marquis of Worcester*, 1865,
p. 563.

used, merely that the engine is "no Pump or Force now in use nor working by any Suckers Barrells or Bellows heretofore used".

Here we leave the Marquis, with the uneasy feeling that we have devoted more space to him than his achievement, whatever it was, deserves. His apparatus may not have been a steam engine at all and, if so, he should not have appeared in this history.

There was another young man in London in the later days of the Marquis, who was attracted to the steam engine, and that was Samuel Morland (1625–95), subsequently created knight and baronet and appointed "Master of Mechanicks" by Charles II. Morland is best known in connection with pumps, especially as the inventor of the packed plunger pump,[1] but his name is associated with several other inventions.

In 1682 we learn that "Sir Samuel Morland has lately shown the King a plain proof of two several and distinct trials of a new invention, for raising any quantity of water to any height by the help of fire alone".[2] What the idea was we do not know with certainty, but research has been made into the matter and from this it appears that it was yet one more attempt to use the pressure of steam to do mechanical work.[3] Nothing came of it, however, as of other ideas such as that of R. D'Acres.[4]

In conclusion, what we have set out above goes to show that the problem of raising water was engaging the attention of many of the acute minds of the day. In support of this statement we may mention that of the fifty-five patents for inventions granted during the reign of Elizabeth, 1561–99, one in seven is for the raising of water, and of the 127 similar patents granted between 1617 and 1642, the same proportion is observable. Only a few, of course, of these had in view the employment of fire for the purpose, but where fire *is* mentioned, except in the case of Papin, it was the pressure of steam that was to do the work.

---

[1] The plunger pump, without packing, was known to the Romans.
[2] *Cal. State Papers, Dom.* Car. II.
[3] Rhys Jenkins, *Collected Papers*, 1936, p. 43.
[4] *Elements of Water-Drawing*, 1659 (Newcomen Society Reprint), p. 9.

It had yet to be realized that the line of approach was not quite so straightforward as that. In evolution the quickest way may be the longest way round, and so it was in the case of the steam engine. It was through the vacuum engine that the conquest was to be made and more than a century was to elapse with steadily advancing technique before this was achieved. At this stage one philosopher—Papin—had struck out on the right path but had faltered by the way, while the practical men were hot foot on a path which, while promising, brought them up against difficulties that they could not overcome in the then state of the mechanic arts.

CHAPTER II

SAVERY AND HIS FIRE ENGINE

Savery employs pressure and suction for raising water—His patent and its extension—Great promise but failure for mining—Used for domestic supply —Papin makes a Savery engine—Blakey makes an automatic one.

IN the preceding chapter we have seen that many minds, scientific and practical, during the seventeenth century, were preoccupied with the problem of finding improved means of raising water from mines, and to a less extent for domestic purposes and irrigation. We have seen also that the field of research for such a power had been narrowed down to two avenues: the combustion of fuel to raise steam from water to a pressure above that of the atmosphere; and the explosion of small quantities of gunpowder to obtain like pressures. The problem of keeping such resultant high pressures within bounds, and the yet greater problem of repeating the operation as often as required with ease and economy, was as far as ever from being solved. It was the first of these avenues that offered the greatest hope of success, and it was this one that was followed; the second avenue was thus neglected, perhaps wisely, for it presented far greater difficulties than the first, and it took over a century before the pioneers cleared the way for the explosion, or as we now term it the internal combustion, engine. The scope of the present volume will not admit of following up the history of the latter branch of the heat engine.

The actual steps in the solution of the problem stated above were taken by Thomas Savery (1650?–1715), who was not only the first to produce a workable apparatus for raising water but was also the first to supply such apparatus to the public. The principle upon which Savery's apparatus worked can be stated briefly. Water is drawn up by suction from the height of, say, 26–28 ft., i.e. that due to the pressure of the atmosphere, by the condensation of steam in a closed vessel; the column of water is there retained, and further steam is used to force the water to a still higher level, suitable cocks turned by hand being provided for the purpose. It is to be observed that the apparatus is partly an atmospheric and partly an above-the-atmospheric pressure apparatus.

We are far from knowing as much as we should like to do about Savery personally. He was the younger son of a family of that name seated at Shilston, in the parish of Modbury, co. Devon, but whether born there or not, and when, is unknown. He is usually given the title of captain, but there is no record of a commission for a person of that name in any English regiment at that time. A Thomas Savery held a commission as "Ensign" in the army between 1688–97, but he cannot be identified with our Savery. It has been stated that he was a military engineer, but there was no such recognized corps in the army at that date, although persons were attached to line regiments for entrenching and such-like duties. Such might have been Savery's status, and some support is lent to this supposed military part of his career by the fact that in 1705 he published a *Treatise on Fortifications,* translated from the Dutch of Baron Coehoorn. A person who had been employed as "trench-master" in the army might well have received the courtesy title of captain on retirement. Another supposition is that Savery was a mine captain in Cornwall, where the title "captain" is given to mine managers, both underground and surface, but there is no evidence that he ever had such employment, while at the same time it is inherently improbable, from what we know about him otherwise. Then again it is suggested that he was a sea captain, but beyond the fact that he invented a ship's log in 1710, and that he wrote a pamphlet entitled *Navigation Improved,* 1698—wherein, however, he disclaims knowledge of maritime affairs—we have nothing upon which to base such a supposition. We should have liked, too, to have known something about Savery's appearance, but the portrait of him which is believed to have once existed is not now to be found.

Savery must have been in favour at Court, for in 1705 he was appointed, by the good offices it is said of Frederick Prince of

Wales, Treasurer to the Sick and Wounded Seamen broken in the Dutch Wars, a post that must have taken up much of his time, and one that he did not relinquish till 1714, when he was appointed Surveyor of the Water Works at Hampton Court. He was elected a Fellow of the Royal Society in 1705.

What is certain is that Savery was, to judge by the evidence of the Patent Rolls, the most prolific inventor of his day; no less than seven patents, or applications for such, between 1694 and 1710, stand in his name. The only one of Savery's patents that is of importance for the present enquiry, however, is that for a "new Invention for Raiseing of Water and occasioning Motion to all Sorts of Mill Work by the Impellent Force of Fire, which will be of great vse and Advantage for Drayning Mines, Serveing Towns with Water, and for the Working of all Sorts of Mills where they have not the Benefitt of Water nor constant Windes", dated July 25, 1698 (No. 356), and covering England, Wales and Berwick-on-Tweed. The grant was for the usual period of fourteen years, but in the following year an Act of Parliament (10 & 11 William III, cap. 61) extended the period of protection for twenty-one years or a total of thirty-five years in all from the date of the original grant. The Act recites the terms of this grant, and gives this as the reason for its extension:

And whereas the said Thomas Savery hath by his great care, paines and expence since the granting the said Letters Patents greatly improved the said Invention...but may and probably will require many yeares' time and much greater expence than hitherto hath been to bring the same to full perfection, and the said Thomas Savery thereby deserves a better encouragement by having a longer terme of yeares allowed him for the sole use and benefitt of the said Invention....

And so on with legal prolixity.

Protection for the invention was extended to Scotland in 1701 by an Act of the Scottish Parliament, granting to James Smith of Whitehill in the county of Midlothian the sole right to use "an engyne or invention for raiseing of water and occasioning motion of millwork by the force of fire". Smith's petition set out what had taken place in England, viz. "that by Articles of

Agreement signed betwixt the said Captain Thomas Savery and the petitioner...he hath allowed and given to the petitioner the communication of the said invention with the model of the said engine to be made use of by him within this Kingdom upon the conditions and provisions contained in the said articles". The protection was to terminate at the same date as that of the extended patent for England.

It is pertinent to enquire how and why Savery turned his attention to steam; all we know on this subject is what he tells us in his own words taken from the dedication of his work: *The Miner's Friend*[1] to the "Gentlemen Adventurers in the Mines of England". It reads thus:

And though my thoughts have been long imployed about water-works, I should never have pretended to any invention of that kind had I not happily found out this new, but yet a much stronger and cheaper form or cause of motion than any before made use of. But finding this of rarefaction by fire, the consideration of the difficulties the miners and colliers labour under by the frequent disorders, cumber-somness, and in general of water-engines, incouraged me to invent engines to work by this new force, that tho' I was obliged to incounter the oddest and almost insuperable difficulties, I spared neither time, pains, nor money till I had absolutely conquer'd them.

We would give a great deal to know how he came by his knowledge of the "rarefaction by fire", and what were the "almost insuperable difficulties" that he overcame; we should like too, to know what means he had in view of "occasioning motion of millwork"; doubtless it was by pumping water back on to a waterwheel from the tail race of the wheel, as he hints in the pamphlet in question. It is true we have some gossip as to the genesis of Savery's invention, gossip that was current at a very early date (see p. 52). This then is the story as retailed by Dr John Theophilus Desaguliers (1683–1744):[2]

CAPTAIN *Savery*, having read the Marquis of *Worcester's* Book, was the first who put in Practice the raising Water by Fire, which he

[1] *The Miner's Friend, or an Engine to raise Water by Fire described, and the manner of fixing it in Mines, with an account of the several other uses it is applicable unto; and an answer to the objections made against it*, by Tho. Savery, Gent., London, 1702.　　[2] *Experimental Philosophy*, II, 1744, p. 465.

proposed for the draining of Mines. His Engine is describ'd in *Harris*'s Lexicon (see the Word *Engine*) which being compared with the Marquis of *Worcester*'s Description, will easily appear to have been taken from him; tho' Captain *Savery* denied it, and the better to conceal the Matter, bought up all the Marquis of *Worcester*'s Books that he could purchase in *Pater-Noster-Row*, and elsewhere, and burn'd them in the Presence of the Gentleman his Friend, who told me this. He said that he found out the Power of Steam by Chance, and invented the following Story to persuade People to believe it, *viz.* that having drank a Flask of *Florence* at a Tavern, and thrown the empty Flask upon the Fire, he call'd for a Bason of Water to wash his Hands, and perceiving that the little Wine left in the Flask had filled up the Flask with Steam, he took the Flask by the Neck and plunged the Mouth of it under the Surface of the Water in the Bason, and the Water of the Bason was immediately driven up into the Flask by the Pressure of the Air. Now he never made such an Experiment then, nor designedly afterwards, which I thus prove:

I MADE the Experiment purposely with about half a Glass of Wine left in a Flask, which I laid upon the Fire till it boil'd into Steam: then putting on a thick Glove to prevent the Neck of the Flask from burning me, I plung'd the Mouth of the Flask under the Water that fill'd a Bason; but the pressure of the Atmosphere was so strong, that it beat the Flask out of my hand with Violence, and threw it up to the Cieling. As this must also have happened to Captain *Savery*, if ever he had made the Experiment, he would not have fail'd to have told such a remarkable Incident, which would have embellish'd his Story.

The story of Savery burning all the copies he could find of a book thirty-five years after its publication sounds rather far-fetched. There is no improbability, however, in the story of the wine flask, in spite of Dr Desaguliers having repeated the experiment with different results. The doctor, we regret to say, seems to have been only too ready to detract from the performances of others than himself, and to attribute bad motives to them, as we shall see again later.

Seeing that we have little more than gossip as to the stages gone through in the development of the invention, and though it would illuminate our minds to know the truth of the matter, we must pass on to the practical exemplification of Savery's

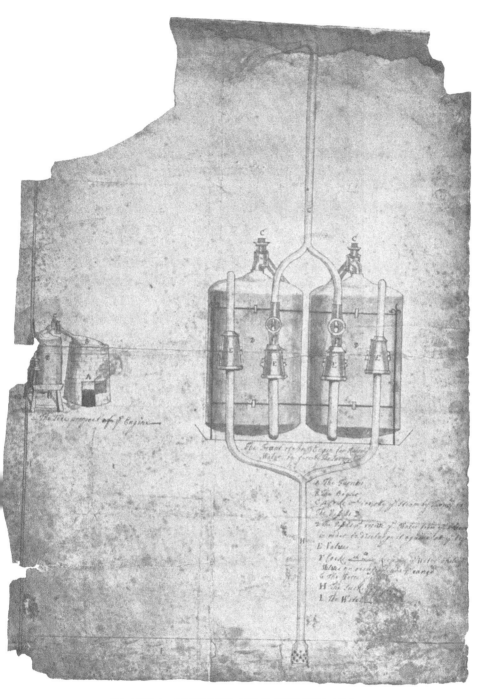

PLATE I. SAVERY'S FIRE ENGINE, 1699

From the drawing in the possession of the Royal Society

ideas in his engine. Patentees, at the time of which we are writing, were not required to file specifications of their inventions, and Savery was no exception to this practice. However, he made a working model of his engine, which he exhibited to the King at Hampton Court Palace—incidentally, this shows that he, as we have said, had access at court; a little later, viz. on June 14, 1699, he showed the model to the Fellows of the Royal Society. The following is the minute of the Meeting: "Mr. Savery Entertained the Royal Society with shewing a small model of his engine for raising water by the help of fire, which he set to work before them; the experiment succeeded according to expectation, and to their satisfaction." This, as Savery says, was "its first appearance in the world". On this memorable occasion, besides demonstrating the model to the Fellows present, Savery furnished the Society with a drawing of his engine; fortunately this has been preserved (see Pl. I). As the legend on it is difficult to read we give the transcript, as follows:

The Front of An Engin for Raising Water by fire by Tho: Savery. The side prospect of ye Engine:

A. The Furniss.
B. The Boyler.
C. 2 Cocks wch convey ye Steam by Turns to The Vessells *D.*
D. The Vessells wch receive ye Water from ye bottom in order to discharge it againe att ye top.
E. Valves.
F. Cocks wch may Keep up ye Water while ye Valves on occasion are Cleans'd.
G. The Force pipe.
H. The Sucking pipe.
I. The Water.

It is from this drawing that the engraving in the *Philosophical Transactions*[1] was prepared, where the account of the engine appears.

It will be observed that there are two receivers, one of which is filled by suction while the other is emptied by pressure, and

[1] *Phil. Trans.* 1699, xxi, No. 253, p. 228.

so on, thereby ensuring a nearly continuous discharge of water. The engine is supported by an ornamental iron frame and stands on a tray on a stool. The boiler seems to be little more than an alembic and no chimney is shown to the fireplace.

In the three years next ensuing, the apparatus underwent considerable improvement in detail, as is shown by the illustration and description in *The Miner's Friend* already referred to, viz. the addition of a hand gear for operating the two steam cocks together; the provision of non-return valves to the receivers which saved perhaps a foot in height; the installation of an auxiliary boiler to replenish the main one and the supplement of a cistern to supply the condensing water.

Savery must have formed very high hopes of the success of his engines, for he set up a workshop for their production, as the following advertisement shows:[1]

Captain Savery's Engines which raise Water by the force of Fire in any reasonable quantities and to any height being now brought to perfection and ready for publick use, These are to give notice to all Proprietors of Mines and Collieries which are incumbred with Water, that they may be furnished with Engines to drain the same, at his Workhouse in Salisbury Court, London, against the Old Playhouse, where it may be seen working on Wednesdays and Saturdays in every week from 3 to 6 in the afternoon, where they may be satisfied of the performance thereof, with less expense than any other force of Horse or Hands, and less subject to repair.

Salisbury Court extends from Fleet Street towards the river Thames, near St Bride's Church. This was the first workshop established in the world for making steam engines, and the advertisement is the first to announce such manufacture commercially.

The better to understand the principle and mode of action of Savery's engine, it will be preferable to consider its early and simple form rather than the one shown in *The Miner's Friend*. Fortunately, we have such an example described and illustrated by Richard Bradley, F.R.S.[2] (Fig. 6). This engine was erected

---

[1] *Post Man*, March 19–21, 1702.

[2] *New Improvements of Planting and Gardening*, 2nd ed. 1718, p. 175.

apparently by Savery himself or his workmen in or about 1712, at Campden House, Kensington, to raise water for domestic purposes. Bradley regarded it as the "truest proportioned of any about London", and it "succeeded so well that there has not been any want of water since it has been built". Refer-ring to the illustration it will be seen that the apparatus com-prised a globular boiler B, with a steam pipe D, in which was a plug stopcock C, supplying steam to a bottle-shaped receiver E, in one with a flat box F. The latter had a pipe G, dipping into the well or water supply below where there was a non-return valve. The box had a rising main L also with a non-return valve, and it delivered by a launder O to a tank or reservoir. Steam was admitted to the receiver and when it was quite hot the stop-cock C was shut, and water from the cock M was allowed to pour

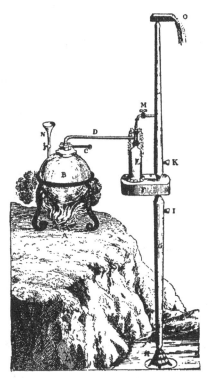

Fig. 6. Savery's fire engine at Kensington, 1712.

From Bradley's *New Improvements of Planting and Gardening*, 1718.

over the outside. The steam inside was condensed, and the atmosphere forced the water up the main G to fill the receiver. Steam was admitted over the surface of the water which was thus forced up the rising main L, where it was again retained by a non-return valve. The cycle could be repeated four times per minute by turning the appropriate cocks; as the receiver held 13 gallons of water, and as the height of the suction pipe was 16 ft. and of the rising main 42 ft., the useful work done by the engine was about 1 h.p. The prime cost of the apparatus was £50.

It will be observed that the actual total lift of the engine was

of the order of 50 ft., so that for the purpose for which the engine
was primarily intended—that of raising water from mines—
there would have to be at intervals of 50 ft. vertically in the pit
a series of engines each delivering into the sump of the one
above. This meant a boiler and engine, sat at every 8 fathoms
of the depth—in fact, such a multiplicity of plant—for mines
had already reached, exceptionally, a depth of 50–60 fathoms—
that the apparatus was hardly to be thought of; in fact, we do
not know with certainty that any plant was ever installed in a
mine, even in a shallow one. What the plant was like that he
intended to put down in a mine is shown in the illustration taken
from *The Miner's Friend*. Bearing in mind the description already
given, it will be observed that he had, by this date, made several
improvements; the vessels or receivers are in duplicate, such
that by acting alternately they gave a greater discharge at a
more even rate than did a single receiver. A second boiler is
provided to heat the water to supply the second. May we not
look upon this as the first boiler feed-water heater? The cocks
for controlling the operations are brought within the convenient
reach of one man. Nevertheless, there was no essential difference
between this apparatus and the one first brought out. We know
that one of these engines was installed at York Buildings in
the Strand, for supplying the neighbourhood with a domestic
supply from the Thames, but even this "was liable to so many
disorders if a single mistake happened in the working of it, that
at length it was looked upon as a useless piece of work and
rejected".[1]

'l'o sum up, the engine found a limited application where the
suction lift was about 20 ft. and the forcing lift somewhat of the
same order, as in the case of supplying water to a mansion, but
for the purpose of raising water from mines, it could not, and
did not, fulfil the sanguine hopes held out for it. Savery dropped
it incontinently after a few years' experience, possibly at the
time when he was appointed Treasurer to the Sick and Wounded,
i.e. in 1705.

[1] Bradley, Richard, *Ten Practical Discourses concerning Earth and Water*,
1727, p. 33.

Desaguliers puts the matter in a nutshell when he says:[1]

CAPTAIN *Savery* made a great many Experiments to bring this Machine to Perfection, and did erect several, which rais'd Water very well for Gentlemen's Seats; but could not succeed for Mines, or supplying Towns, where the Water was to be rais'd very high and in great Quantities: for then the Steam requir'd being boil'd up to such a Strength, as to be ready to tear all the Vessels to pieces. . . . I have known Captain *Savery*, at *York*-Buildings make steam eight or ten times stronger than common air (i.e. 117 to 147 lb. per sq. in.); and then its Heat was so great that it would melt common soft Solder; and its Strength so great as to blow open several of the Joints of his Machine; so that he was forc'd to be at the Pains and Charge to have all his Joints solder'd with Spelter or hard Solder.

If true this would certainly be the case as the temperature of steam of ten atmospheres pressure is 357° F. and soft solder melts below that.

Some reader may remark that it would have been easy to increase the height of lift by increasing the boiler pressure; so it would were it not that the state of the constructional arts of the time was not sufficiently advanced to admit of making boilers to stand the pressure so required. The failure of the engine must be attributed in the main to imperfections of workmanship, and the unreliability of materials. To-day there would be no difficulty in making a successful engine to Savery's drawings, indeed, in a modified form and with an automatic valve, Savery's principle is embodied in that most useful emergency pump, Hall's pulsometer of 1876, still manufactured extensively.

It is curious, by the way, to observe what an attraction Savery's idea has had, for one inventor after another has dabbled in it. We must mention Papin, who turned aside in 1707 from the true line of advance, i.e. the cylinder-and-piston engine, to make an engine on Savery's plan. Papin, who was then at Hesse-Cassel, submitted his proposal to the Royal Society with a request for a grant of money to carry out experiments upon it. The Society naturally referred the proposal for report to Savery, their member most qualified to do so. He found nothing in it

[1] *Experimental Philosophy*, II, 1744, p. 466.

that he had not already done himself, and did not recommend the grant. In the course of his criticism, there is a noteworthy passage in which he goes out of his way to damn the cylinder-and-piston engine; he considered it was impossible to make it work, because the friction set up would be too great! This is significant as we shall see when we come to Newcomen.

The only other inventor of the Savery type engine that we need mention is William Blakey, because he went further than the rest and made his design of the engine fully automatic. This was in 1776,[1] but alas he was at the heel of the hunt, for long before this time a safe and fairly efficient engine for raising water had been invented, and to this we must devote the next chapter.

[1] *Trans. Newcomen Soc.* XVI, p. 97.

CHAPTER III

NEWCOMEN AND HIS VACUUM ENGINE

Newcomen's engine as early as Savery's but fundamentally different—
Mårten Triewald's story—Early difficulties surmounted—Detractors—
Dudley Castle engine—Death of Savery—His patent acquired by certain
"Proprietors"—Newcomen checked—Expiration of Savery's patent—
Spread of the engine in Great Britain and on the Continent—Death of
Newcomen.

DURING the time that Savery's invention, with the brilliance of a meteor, had appeared on the scene and then tailed off into obscurity, another man of Devon was wrestling with the same problem of raising water by fire, and was solving it in an entirely different manner to that of Savery. This man was Thomas Newcomen, the inventor and begetter of the atmospheric engine, whence the steam engine of the present day can be traced in uninterrupted descent—the man who made the first and greatest step in its development, and "it is the first step that costs". Unfortunately, it is a development of which we know very little.

To put it briefly, Newcomen's apparatus consisted of a cylinder provided with a piston, below which there was access of steam from a separate boiler, and a cock by which a jet of cold water could be turned on to condense the steam. When this was done, the pressure of the atmosphere forced down the piston, and by suspending the latter from the end of a lever or beam, which acted as a huge pump-handle, water was raised by a pump-buckct attached to the other end of the beam (see Fig. 7). At the time when, as Savery had found to his discomfiture, the art of making a vessel to resist internal pressure was scarcely understood, it was vital to success that Newcomen did not need to employ steam of pressure higher than that of the atmosphere. When we look into the matter closely, the extraordinary fact emerges that the new engine was little more than a combination of known parts; a cylinder and a piston had been used by

Guericke, Papin and others; the pump-bucket and the pump-handle were well known, although not on the scale now contemplated; the boiler with its setting was little more than a

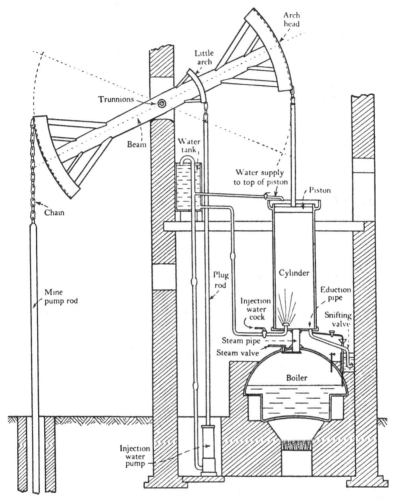

Fig. 7. Diagram of Newcomen's atmospheric engine, 1712.

large brewer's copper; the layer of water on the piston to act as packing was a novelty but not a strikingly original idea. On the other hand the jet of water to condense the steam inside the cylinder was a new and important invention. Such was Newcomen's simple but brilliant combination.

It will be easier to understand how so great an invention came to fruition if we glance at what has been gleaned about the man himself, and relate what we have been able to gather about the tribulations through which he must have passed, for he soon found, as all inventors do, that there were many pitfalls to be overcome before he could reduce his ideas to practice.

Although Newcomen was a contemporary, nay almost a neighbour of Savery, there is no hint that they were acquainted, or that they knew of one another's experiments. Newcomen was born at Dartmouth, co. Devon, and christened on February 28, 1663. He was the great-grandson of the Rev. Elias Newcomen (1547–1614), rector of Stoke Fleming, a village quite close to Dartmouth, and a member of a family that had been of consequence since the twelfth century,[1] established at Saltfleetby in Lincolnshire. This is mentioned merely to suggest that Newcomen came of good stock, and was a man of whose mental powers something might be expected. We know nothing of his schooling—most likely it was of little account, as was so usual at that date—nor yet of his apprenticeship to the trade of an ironmonger, believed to have been served in Exeter. We are to imagine him established in business for himself in his native town certainly before 1703, for in that year we find his name in the mayoral accounts as supplying nails for the church. He married in 1705 Hannah Waymouth of Malborough, co. Devon. It is frequently asked whether any portrait or description of his appearance is known, and we are obliged to reply that none is extant.

How and why it was that Newcomen turned his attention to steam and its possibilities may be gathered from the statements of contemporaries. Mårten Triewald (1691–1747), a Swede, came to this country in 1716 to study natural science and mechanics, and thereby got into touch in London with Dr Desaguliers, already mentioned, who was the best known teacher of his day. Through him Triewald made several acquaintances, and found an opening for his talents in the new business of erecting engines. He allows us to infer that he knew Newcomen personally, so

[1] *Trans. Newcomen Soc.* IX, p. 105.

that much weight must be given to Triewald's statement which is as follows:[1]

Now it happened that a man from Dartmouth, named Thomas Newcomen, without any knowledge whatever of the speculations of Captain Savery, had at the same time also made up his mind, in conjunction with his assistant, a plumber by the name of Calley, to invent a fire-machine for drawing water from the mines. He was induced to undertake this by considering the heavy costs of lifting water by means of horses, which Mr. Newcomen found existing in the English tin mines. These mines Mr. Newcomen often visited in the capacity of a dealer in iron tools with which he used to furnish many of the tin mines.

The statement that Newcomen had no knowledge of what Savery was doing, is corroborated by another writer, Stephen Switzer, who says:[2]

I am well inform'd that Mr. *Newcomen* was as early in his Invention, as Mr. *Savery* was in his, only the latter being nearer the Court, had obtain'd his Patent before the other knew it, on which account Mr. *Newcomen* was glad to come in as a Partner to it.

The inference from this statement is that Newcomen had begun his speculations and made his attempts to construct a steam engine at least as early as 1698, the date of Savery's patent. Triewald's statement that "for ten consecutive years Mr. Newcomen worked at this fire-machine" can be accepted unhesitatingly as true, indeed, as we shall see later, the period must actually have been longer than this.

Dr John Robison (1739–1805), made the statement in 1797 that Newcomen was acquainted with Robert Hooke (1635–1703), sometime Secretary of the Royal Society, and obtained from the latter notes on the work of Papin, and suggestions as to how to proceed with his invention. Repeated search has failed to find any documents to substantiate what seems, on the face of it, an inherently improbable story. There is the distinct possibility, however, that Newcomen had seen in the *Philosophical Transactions* of the Royal Society for 1697 the review

---

[1] *Kort Beskrifning om Eld- och Luft-Machin*, 1734. Engl. trans. 1928.
[2] Switzer, *Hydrostaticks and Hydraulicks*, 1729, p. 342.

that appeared there of the article of Papin of 1690 describing his cylinder-and-piston engine already referred to on p. 11, but even this is to suppose that an ironmonger in a provincial town knew of what was going on in the scientific world of London. The conclusion now reached is that Robison's statement is unworthy of credence,[1] but it has been repeated so often that we have felt bound to take cognizance of it.

What were the contributions of Newcomen and his assistant Calley respectively to the solution of the problem we do not know, but we may hazard a guess that Newcomen supplied the brains and Calley most of the technical skill. Of the difficulties, unforeseen and harassing, that Newcomen and Calley had to overcome in these long years of preparation, we are in complete ignorance. First of all they could only work on a model scale, in which extraneous effects such as friction bear a far higher proportion to the total effect than in an actual engine. To get a cylinder of any greater diameter than about 7 in., the size that was then ordinarily made for pumps, was a difficulty, and to get it bored truly cylindrical was beyond the capacity of the pump-makers and gun-founders of the day, who alone could undertake such work. It seems that Newcomen had to be content with a cast brass cylinder laboriously rubbed smooth on the inside with sand and elbow grease. The piston consequently could only be a rough fit in the cylinder, and to meet the difficulty of making it tight, packing had to be devised, a problem that has proved a perennial one for engineers. Newcomen solved the problem for his own engine by adopting a leather flap around the edge of the piston, with water on the top to seal the leather.

To condense the steam it is quite probable that, in the early stages as Desaguliers states, cold water was poured over the outside of the cylinder, but the superior method of applying a jet of water inside the cylinder must have early suggested itself. The story that this method of condensation was discovered by the accident of a hole occurring in the cylinder or in the piston is inherently improbable, as such a hole would let in water or air continuously and thus effectively prevent any vacuum at all

[1] *Trans. Newcomen Soc.* XVII, p. 6.

being formed. The water used to condense the steam and the condensate itself were to be got out of the cylinder. This was effected by a "sinking" or eduction pipe fixed to the bottom of the cylinder, and extended downwards a few feet vertically to enable the water to run off by gravity on readmission of steam. The mouth of this pipe was upturned and covered by a water-sealed flap valve which prevented water running back when the vacuum was re-formed.

To transmit the power to the pump-handle or beam, which required to oscillate on trunnions, arch heads at both ends and chain connections to the piston and pump rods respectively had to be schemed; probably this did not require much thought, but it is well to remember that there were few books available, and possibly none within Newcomen's reach, to which one could refer for hints as to such mechanical contrivances.

Then there was the valve to turn on the steam and the cock to turn on the condensing jet, each to be opened and shut at the right moment. No doubt, at first, Newcomen would have some person present to work the valves by hand, but this was obviously undesirable as a permanent arrangement, and soon he contrived a way of effecting the movement of the valves by the motion of the beam itself. This alone was a capital invention for, with the exception of the clock, there was no example then extant of a mechanism controlling its own movements, or in a word, being self-acting. Of all the contrivances invented by Newcomen, this was the most ingenious, and most influenced subsequent invention.

When at last Newcomen had got the model to work, he must have suffered intense disappointment; the engine would go merrily for a time, and then gradually slow down until at last it would stop. Here was a poser! It is unlikely that he would have found out at once the reason for this obscure phenomenon, arising as it does from the fact that atmospheric air, always found dissolved in water to a small amount, is given off with the steam when water is boiled, and not being condensable, accumulates in the cylinder, giving rise to what is known as "wind-logging". It is probable that Newcomen found out by trial the remedy which was to let the steam flow through his cylinder after each stroke for a few seconds, carrying the air with it. To make quite sure that the air was removed, he contrived a "snifting" valve, so named from the noise it made, situated in a basin or cistern of water on the eduction pipe. By observing that the air bubbled out there at every steam stroke, one could be sure that the cylinder had been freed from air.

The construction of the boiler cannot have presented much difficulty. It would appear as if the inventors had borrowed ideas, as far as construction and setting, from the brewer's copper, but nevertheless, provision had to be made for feeding the boiler with water, which was done by a standpipe, and for means to show the height of the water within it, which was done by a couple of gauge pipes.

Something should be said about the materials then available; it needs no stretch of imagination to suppose that chains would break, pipes would burst, leather would tear away and incrustation would form in the boiler and on the interior of the cylinder. Who sustained these men's spirits during the ten or more years of experiment before success was attained? Who supplied the funds for carrying on the work? The ironmongery business must have suffered from neglect unless we conjecture that Newcomen's wife Hannah carried it on; certainly, since we do find her doing so at a later period when her husband was away on engine erection, we do well to remember Hannah as the staunch helpmate.

Perhaps we are mistaken in supposing that the series of inventions outlined above required all the acumen with which we, basing our opinion upon what we should have had to go through ourselves had we like problems to solve and were capable of solving them, have credited the partners. Might not rather a chapter of accidents have led to the happy result? Listen to what our esteemed mentor, Dr Desaguliers, whom we have already mentioned, and who was somewhat of an inventor himself, has to say on the subject:[1] "If the Reader is not acquainted with the History of the several Improvements

---

[1] *Experimental Philosophy*, II, 1744, p. 474.

of the Fire-Engine since Mr. *Newcomen* and Mr. *Cawley* first made it go with a Piston, he will imagine that it must be owing to great Sagacity, and a thorough Knowledge of Philosophy, that such proper Remedies for the Inconveniences and difficult Cases mention'd were thought of: But here has been no such thing; almost every Improvement has been owing to Chance." Delicious! and if true quite contrary to the course of events in the case of every other capital invention. In our reading of that history, nothing has ever "been brought to bear" without, as the older patent specifications express it, "great labour, pains and expence".

But to continue our story. Here for lack of better evidence we have to rely on Desaguliers's account,[1] and this is what he says:

About the year 1710, *Tho. Newcomen*, Ironmonger, and John Calley, Glazier, of *Dartmouth* in the *County* of *Southampton* [he means Devonshire], (Anabaptists), made then several Experiments in private, and having brought it to work with a piston, &c. in the latter End of the Year 1711, made Proposals to draw the Water at *Griff* in *Warwickshire*; but their Invention, meeting not with Reception, in *March* following, thro' the Acquaintance of Mr. *Potter* of *Bromsgrove* in *Worcestershire*, they bargain'd to draw water for Mr. *Back* of *Wolverhampton*, where, after a great many laborious Attempts, they did make the Engine work; but not being either Philosophers to understand the Reasons, or Mathematicians enough to calculate the Powers, and to proportion the Parts, very luckily by Accident found what they sought for. They were at a loss about the Pumps, but being so near *Birmingham*, and having the Assistance of so many admirable and ingenious Workmen, they soon came to the Method of making the Pump-Valves, Clacks and Buckets; whereas they had but an imperfect Notion of them before.

Desaguliers seems to take it almost as an affront that anyone not a philosopher or mathematician should succeed in making a fire engine, and his account reads as if Newcomen and Calley had found their paths strewn with roses, in spite of their crass ignorance! Search has been made without avail to find particulars of "Mr. Back of Wolverhampton" and of the engine the inventors set up for him. We are at a loss, unless the latter was

---

[1] *Experimental Philosophy*, ii, 1744, p. 532.

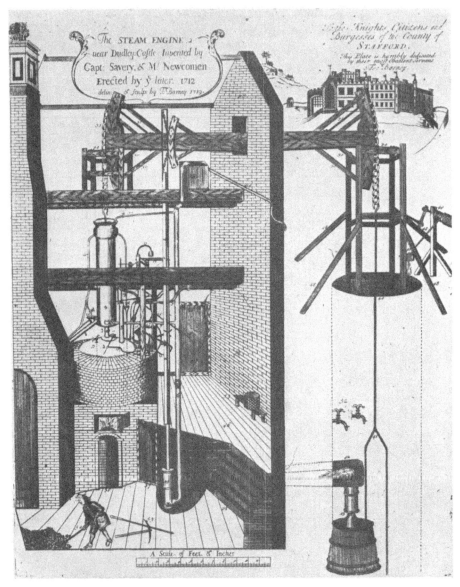

PLATE II. THE STEAM ENGINE NEAR DUDLEY CASTLE, STAFFORDSHIRE, 1712

From the engraving by T. Barney, 1719

" *The STEAM ENGINE near Dudley Castle Invented* by Capt. Savery, & Mr Newcomen Erected by ye *later*, 1712", an engraving of which was executed in 1719 by Thomas Barney, and of which several copies have come to light during the last sixty years (see Pl. II). The correspondence in point of date and the fact that Dudley Castle is not far from Wolverhampton renders it a matter almost beyond question that the engine mentioned by Desaguliers and the engine shown in the engraving are one and the same. From the testimony of Triewald we have corroboration of the date of the erection of the first engine in these words: "Mr. Newcomen erected the first fire-machine in England in the year 1712, which erection took place at Dudley Castle in Staffordshire." A point of great interest is that the engine is described on the engraving for the first time as a "Steam Engine". On the same sheet as the engraving is a letterpress description and as this is the first extant description of a steam engine, we venture to give it in full thus:

REFERENCES

By Figures,

To the several Members

1 | The Fire Mouth under the Boyler with a Lid or Door.
2 | The Boylor 5 Feet 6 Inches Diameter, 6 Feet 1 Inch high, the Cylindrical Part 4 Feet 4 Inches, Content near 13 Hogsheads.
3 | The Neck or Throat betwixt the Boyler and the Great Cylinder.
4 | A Brass Cylinder 7 Feet 10 Inches high, 21 Inches Diameter, to Rarifie and Condense the Steam.
5 | The Pipe which contains the Buoy, 4 Inches Diameter.
6 | The Master Pipe that Supplies all the Offices, 4 Inches Diameter.
7 | The Injecting Pipe, fill'd by the Master Pipe 6, and stopp'd by a Valve.
8 | The Sinking Pipe, 4 Inches Diameter, that carries off the hot Water or Steam.
9 | A Replenishing Pipe to the Boyler as it wastes, with a Cock.
10 | A Large Pipe with a Valve to carry the Steam out of Door.
11 | The Regulator moved by the 2 Y y and they by the Beam 12.
12 | The Sliding Beam mov'd by the little Arch of the Great Beam.
13 | Scoggen and his Mate who work double to the Boy, Υ is the Axis of him.
14 | The Great Y that moves the little y and Regulator, 15 and 11 by the Beam 12.
15 | The little y, guided by a Rod of Iron from the Regulator.

16 | The Injecting Hammer or F that moves upon its Axis in the Barge 17.
17 | Which Barge has a leaking Pipe, besides the Valve nam'd in Nº 7.
18 | The Leaking Pipe 1 Inch Diameter, the Water falls into the Well.
19 | A Snifting Bason with a Cock, to fill or cover the Air Valve with Water.
20 | The Waste Pipe that carries off the Water from the Piston.
21 | A Pipe which covers the Piston with a Cock.
22 | The Great Sommers that Support the House and Engine.
23 | A Lead Cystern, 2 Feet square, fill'd by the Master Pipe 6.
24 | The Waste Pipe to that Cystern.
25 | The Great Ballanc'd Beam that works the whole Engine.
26 | The Two Arches of the Great Ballanc'd Beam.
27 | Two Wooden Frames to stop the Force of the Great Ballanc'd Beam.
28 | The Little Arch of the Great Ballanc'd Beam that moves the Nº 12.
29 | Two Chains fix'd to the Little Arch, one draws down, the other up.
30 | Stays to the Great Arches of the Ballanc'd Beam.
31 | Strong Barrs of Iron which go through the Arches and secure the Chains.
32 | Large Pins of Iron going through the Arch to stop the Force of the Beam.
33 | Very strong Chains fixed to Piston and the Plugg and both Arches.
34 | Great Springs to stop the Force of the Great Ballanc'd Beam.
35 | The Stair-Case from Bottom to the Top.
36 | The Ash-hole under the Fire, even with the Surface of the Well.
37 | The Door-Case to the Well that receives the Water from the Level.
38 | A Stair-Case from the Fire to the Engine and to the Great Door-Case.
39 | The Gable-End the Great Ballanc'd Beam goes through.
40 | The Colepit-mouth 12 Feet or more above the Level.
41 | The dividing of the Pump work into halves in the Pit.
42 | The Mouth of the Pumps to the Level of the Well.
43 | The Pump-work within the Pit.
44 | A Large Cystern of Wood 25 Yards or half way down the Pit.
45 | The Pump within the House that Furnishes all the Offices with Water.
46 | The Floor over the Well.
47 | The Great Door-Case 6 Feet square, to bring in the Boyler.
48 | Stays to the Great Frame over the Pit.
49 | The Wind to put them down gently or safely.
50 | A Turn-Barrel over the Pit, which the Line goes round, not to slip.
51 | The Gage-Pipe to know the Depth of the Water within the Boyler.

52 | Two Cocks within the Pit to keep the Pump work moist.
53 | A little Bench with a Bass to rest when they are weary.
54 | A Man going to Replenish the Fire.
55 | The Peck-Ax and Proaker.
56 | The Centre or Axis of the Great Ballanc'd Beam.

On one of the engravings of the Dudley Castle engine there is a MS. note in contemporary handwriting which tells us that the beam "Vibrates 12 times in a Minute & each stroke lifts 10 Gall. of water 51 yards p'pendr", i.e. 120 hogsheads per hour and this means that the engine was of about 5½ h.p.

We should like to know what was the subsequent history of this historic engine. We learn from an eye-witness that it was still at work in 1725, throwing up about "60 hogsheads of water in an hour's time";[1] in all probability it worked on until the pit was wrought out.

It would be a matter of unusual interest if the exact site of the engine were known, but unfortunately this is not the case. Basing his examination on geological grounds, and topographical considerations, Mr T. E. Lones has succeeded in narrowing down the possible site to within an area of about 9 acres, between the outcrops of thick coal and the broach coal at Tipton, Staffordshire.[2] The surface here has been so much disturbed by mining operations and by building that nothing can be confidently said about the site.

NEWCOMEN'S ENGINE COMPLETED

When we examine the atmospheric engine as delineated in working order on Barney's engraving, and compare it with the elementary apparatus that had preceded it, we are astonished at the prodigious advances made in comparatively so short a time. We have now before us a thoroughly practical job schemed by the brain of Newcomen and carried into effect by the united efforts of carpenter, plumber, blacksmith, coppersmith, pumpmaker and bricklayer, under the direction of Calley; the remarkable fact is that no demand had to be made upon the craftsmanship

[1] John Kelsall's *Diary*, 21. 6 mo. (i.e. September 21), 1725.
[2] *Trans. Newcomen Soc.* XIII, p. 1.

of any of these trades other than what was made ordinarily every day.

We have already pointed out that one of the most important practical features of the engine was that it was self-acting. At the date of Barney's engraving, 1719, this gear was in use and there is no reason to suppose that it was not on the engine when erected. The motion was derived from a sliding beam or plug-rod, hung from a small arch head on the beam, and reciprocated with it. Pins were so placed in this rod as to strike the levers actuating the steam valve, and the injection cock respectively, at the right moments in the up and down strokes. The steam valve was a simple sector-shaped plate sliding across a corresponding aperture immediately below the cylinder. The injection cock was nothing but a beer-barrel cock such as Savery had used. The latter was actuated by an ingenious device known as the "tumbling bob" on the injection cock; it was a λ-shaped lever centred on the cock; the legs were actuated by the pins on the plug-rod and once past the vertical the bob overbalanced and ensured positive and instantaneous opening or closing, obviating sticking of the injection cock. It is typical of Newcomen's inventiveness that he had an alternative method, earlier in point of date, of working this cock. This consisted of a pipe in the boiler dipping below the water level; in the pipe was a buoy with a projecting stem or rod. A slight rise of pressure in the boiler forced the water up the buoy pipe carrying with it the buoy, and its stem tripped the injection cock by an attached cord. Thus there were two principles of working: one dependent upon, and the other independent of, the pressure of the steam. In one case the speed of the engine was dependent on the rate of firing of the boiler, and in the other case the speed was constant provided steam was kept up. On this point Desaguliers gives us some details:[1]

They used before (i.e. 1713) to work with a Buoy in the Cylinder [he means "boiler" of course] inclos'd in a Pipe, which Buoy rose when the Steam was strong and open'd the Injection, and made a Stroke; thereby they were capable of only giving six, eight, or ten

---

[1] *Experimental Philosophy*, ii, 1744, p. 533.

Strokes in a Minute 'till a Boy, *Humphry Potter*, who attended the Engine, added (what he call'd *Scoggan*) a Catch that the beam Q [he means the "plug-rod"] always open'd: and then it would go 15 or 16 strokes in a Minute. But being perplexed with Catches and Strings, Mr. *Henry Beighton*, in an Engine he had built at *Newcastle* on *Tyne* in 1718, took them all away, the Beam itself simply supplying all much better.

The engine in question was that erected at Washington Colliery.

The story of the boy Potter, particularly as embroidered by subsequent writers by the added statement that he invented the scoggan so that he could run away and play, has always caught the popular fancy; indeed it would hardly be untrue to say that Potter is better known than Newcomen. Even the moralists have somehow found virtue in the young rascal's imputed dereliction of duty. Sad to relate, no confirmation of the story is to be found, and it is most probable that Desaguliers, all of whose information was obtained at second-hand from Henry Beighton, land surveyor and F.R.S. (1686–1743), may not have grasped the two different methods of working the injection cock, and have confused the buoy in the pipe with the boy Potter in the flesh. It is a fact that men of this name were busily erecting engines at the time Desaguliers was writing on the subject, and one of these men may have added this "scoggan".

It is likely that the Dudley Castle engine and its performance would become common talk in mining circles, and there must have been, as we shall see later, a number of persons, undeterred by the cost, who decided to erect engines, but between 1712 and 1715 we have no documentary proof of any. There is another possibility and it is this: it would be natural for the inventors to seek to protect their invention by letters patent. The statement has frequently been made that Newcomen did obtain a patent; so far from this being the case, there is no record that he even made an application for one, and for a very good reason —he would learn on enquiry that he had been anticipated by Savery, and that the latter's patent was still in force. We can imagine Newcomen's chagrin and disappointment! In spite of the fact that the engines were as dissimilar as they could possibly be, there was no denying the fact that Newcomen's engine was just as much worked "by the Impellent Force of Fire" as was that of Savery. In other words, Savery's was a master patent, covering any and every kind of heat engine. The Act of Parliament of 1699 is quite specific on this point: no person but he or his assigns may "make, imitate, use or exercise any vessells or engines for raiseing water or occasioning motion to any sort of millworks by the impellent force of fire". What were Newcomen and Calley to do? Triewald puts the case in a nutshell: "Although the invention was thus completed, the inventors however could not obtain any privilege because Captain Savery already possessed a privilege....for constructing a fire engine....Mr. Newcomen and Mr. Calley did not see any other way out of their difficulties but to join Capt. Savery and form a Company." Switzer corroborates this statement in these words: "Mr. *Newcomen* was glad to come in as a Partner to it" (i.e. the patent). Newcomen was in the strong position of having the knowledge and technique without which the patent privilege was valueless, but Savery or his representatives, backed by the Act of Parliament, could stop Newcomen, and may have done so. The obvious course was for the parties to come to terms. What these terms were and when they were arrived at we do not know, but it is probable that it was about 1715. Savery died in that year and the deal may even have taken place with his executors. However that may be, we find that Savery's privileges were acquired by certain "Proprietors", and we draw this inference from an advertisement in the *London Gazette*, August 11–14, 1716, which reads as follows:

Whereas the Invention for raising Water by the impellant force of Fire, authorized by Parliament, is lately brought to the greatest Perfection; and all sorts of Mines, &c may be thereby drained, and Water raised to any Height with more Ease and less Charge than by the other Methods hitherto used, as is sufficiently demonstrated by diverse Engines of this Invention now at Work in the several Counties of Stafford, Warwick, Cornwall and Flint. These are therefore to give Notice, that if any Person shall be desirous to treat with the Proprietors for such Engines, Attendance will be given for that Purpose every Wednesday at the Sword-Blade Coffee-house in Birchin Lane,

London, from 3 to 5 of the clock; and if any Letters be directed thither to be left for Mr. Elliot, the Parties shall receive all fitting Satisfaction and Dispatch.

It will be observed that neither Savery's nor Newcomen's name is mentioned. Engines are stated to be at work in several counties of England: that in co. Stafford was undoubtedly the Dudley Castle engine; the one in co. Warwick must have been that at Griff Colliery, near Coventry; that in Cornwall was probably the one at Huel Vor, Breage, between Helston and Marazion, and the one in co. Flint was almost certainly that at Hawarden, erected in 1715. As a matter of fact the engine had spread since 1712 more widely than the advertisement indicates. We know of an engine near Whitehaven, Cumberland, one at Tanfield Lea, 1715, near Newcastle-upon-Tyne, and another at Austhorpe, near Leeds in Yorkshire. It would seem as if Newcomen had gone on quietly erecting engines as called for, ignorant of or disregarding Savery's patent, but of this we cannot be sure. The engine at Austhorpe is of particular note because John Smeaton (1724–92), the celebrated engineer, resided close by at Austhorpe Lodge. His attention was thus directed to it, and this resulted in his making enquiries as to its working, with the outcome that we shall see later.

Henry Beighton, whom we have already mentioned, resided at Griff, in Warwickshire, and was the first scientific man to study the performances of the engine, which was situated, so to speak, at his own door. In 1717 he calculated a table showing the powers of the atmospheric engine, and as it is the earliest of its kind, it is here reproduced (see Fig. 8).[1] It is not to be inferred that this represents what was common practice; rather it was intelligent interpolation on Beighton's part. It will be noticed that he anticipated the greatest depth from which water would be drawn was 100 yd., that the largest size of cylinder would be 40 in. diameter, and that the mean effective pressure would be 8 lb. per sq. in. Accompanying his table Beighton

---

[1] Published in the *Ladies' Diary*, 1721, p. 22, of all places. The explanation is that Beighton was its editor from 1713 till his death.

has some good advice to give and a few pertinent remarks to make[1] about engineers:

It were much to be wish'd, they who write on the Mechanical Part of the Subject, would take some little Pains to make themselves Masters of the Philosophical, and Mechanical Laws of (Motion or) Nature; without which, it is Morally impossible, to proportion them [i.e. the Machines for draining of Mines] so as to perform the desired End of such Engines. We generally see, those who pretend to be *Engineers*, have only guess'd, and the Chance is, they sometimes

A *Physico-Mechanical* Calculation of the Power of an *Engine*.

Diamerer of the Pump Inches.	Draws at a 6 Foot stroke.	at 16 St. in a Min. draws per Hour.	The Diameter of the Cylinder. (The Depth to be Drawn in Yards.)												
Inch	Ale Gall.	Hogsh. Gall.	15	20	25	30	35	40	45	50	60	70	80	90	100
4	3. 20	48. 51		Inches			9	10	11	11½	12	13½	14	15	16
4½	4. 04	60. 60				10	11	11¾	12	13	14	15	16	17	18¼
5	5. 02	66. 61			10	11	11¾	13	13¾	14	15½	16¾	18½	19½	20½
5½	6. 26	94. 30		10	11	12	13	14	15	15¾	17	19	20	21	22¼
6	7. 22	110. 1	9½	11	12	13	14	15½	16	17	19	20½	22	23	24½
6½	8. 46	128. 54	10	12	13	14	15½	16¼	18	19	20	22	23	24½	26¼
7	9. 82	149. 40	10¾	13	14	15½	16¼	18¼	19	20½	22	24	25½	27	28½
7½	11. 32	172. 30	11	13¾	15	16½	18	19	20	21¼	23¼	25	27	28½	30¼
7¾	12. 02	182. 12	12	14	15½	17½	18¾	19¾	21	22	24¼	26	28	29	31½
8	12. 82	195. 22	12½	14½	16½	18½	19	20½	21½	23	25	27	29	30½	32½
8½	14. 52	221. 15	13½	15¼	17¼	19	20¼	21¼	23	24	26¾	28½	31	32½	35¼
9	16. 24	247. 7	14	16½	18	20	21½	23	24¼	25	28	30½	33	35	36½
10	20. 04	304. 48	15½	18	20	22	23¾	25¼	27	28¾	31¼	33¾	36	38¼	40

Fig. 8. Table of the powers of the atmospheric engine.
From the *Ladies' Diary*, 1721.

succeed; else they have made them like others that have done pretty well. But he who has skill enough in *Geometry*, to reduce the *Physico-Mechanical* Part to Numbers, when the Quantity of Weight or Motion is given, and the Force designed to move it, can bring forth all the Proportions, in a Numerical Calculation, so as it may be almost impossible to Err.

Beighton was a friend of Desaguliers and communicated to the latter a great deal of practical information, particularly on atmospheric engines, which is retailed by the author as if it were his own. Bearing upon the table just cited there is an

[1] *Ladies' Diary*, 1721, p. 21.

interesting scrap of information that must have come through Beighton and that is as to the way in which Newcomen calculated the power of his engine:[1]

Mr *Newcomen's* Way of finding it was this: From the Diameter [of the cylinder] squar'd he cut off the last Figure, calling the Figure on the left Hand long Hundreds, and writing a Cypher on the right Hand, call'd the Number on that Side, Pounds; and this he reckon'd pretty exact as a Mean, or rather when the Barometer stood at 30, and the Air was heavy. N.B. *This makes between* 11 *and* 12 *Pounds upon every superficial round Inch.* Then he allow'd between $\frac{1}{3}$ and $\frac{1}{4}$ Part for what is lost in the Friction of the several Parts and for Accidents: and this will agree pretty well with the Work at *Griff* Engine, there being lifted at every stroke between $\frac{2}{3}$ and $\frac{3}{4}$ of the weight of the atmospherical Column pressing on the Piston.

This works out at a mean effective pressure of 9·4 lb. but to be on the safe side subsequent engineers reckoned on 7 lb. or half an atmosphere.

The need for unwatering mines economically was almost as pressing on the Continent as in Britain, so that it is not to be wondered at that a knowledge of "the beautifullest and most useful Engine that any Age or Country ever yet produc'd"[2] soon penetrated there. The first engine to be erected was in 1722 at Königsberg, 3 miles from Schemnitz, in the mining district of Upper Hungary, what is now Czechoslovakia. The course of events seems to have come about in this wise: Joseph Emanuel Fischer von Erlach (1693–1742), son of a well-known architect at the Court of Vienna, visited England during his grand tour, and there learnt about the new agent for draining mines, the very thing needed in Hungary. He appears to have taken back with him Isaac Potter, described as an English engineer, who erected the engine in question. A drawing of it made in 1753, now in the Deutsches Museum, Munich, shows, as one would expect, no departure from English practice. Once started, quite a number of similar engines were erected in this district by Potter and he was looked upon there as the inventor.

---

[1] *Experimental Philosophy*, II, 1744, p. 482.
[2] Switzer, *Hydrostaticks and Hydraulicks*, II, p. 335.

He may have been a relation of the Humphry Potter mentioned
above, or of John Potter who was an agent in the North of
England for the erection of fire engines, as the following advertisement from the *Newcastle Courant* of January 27, 1724, shows:
"This is to give notice to all gentlemen and others, who have
occasion for the fire engine, or engines for drawing of water
from the collieries, &c. to apply to Mr. John Potter in Chester-
le-Street, who is empowered by the Proprietors of the said fire
engines to treat about the same." The obvious inference from
this advertisement and from the town which is named in it, is
that Potter had been appointed as agent to work the Durham
and Northumberland coalfield, but strangely enough, although
there were fourteen engines erected in this coalfield before the
date of the expiration of the patent, there is no evidence of any
agreement with Potter or with the "Proprietors", nor of any
royalty payments for the engine.[1] This is all the more strange
because Potter erected engines still further afield, i.e. in Scotland.
Detailed particulars of an engine erected by him at Edmonstone
Colliery, Midlothian, have been preserved. The first document,
dated 1725, is: "A *licence* granted by the *Committee* in London
appointed and authorized by the *Proprietors* of the *Invention* for
raising Water by *Fire* to *ANDREW WAUCHOPE* of *Edmonstone Esquire*." The licence recites that, as the colliery could
not be wrought by reason of water, liberty is granted to erect
one engine with a steam cylinder nine feet long by twenty-eight
inches diameter, according to the method and manner now used
at *Elphingstone* in Scotland; for which licence a royalty of £80
per annum was to be paid for eight years.[2] The wording of this
clause is of particular interest in that the licence is granted to
"Andrew Wauchope, his heirs executors or assignees to hold
use and exerce the samen engine so to be erected from and
after the 24th day of June next ensuing the date hereof" (i.e.
May 1725) "for and during and until the full end and period of
the said John Meres and Proprietors aforesaid their grant and

---

[1] *Trans. Newcomen Soc.* XVII, p. 131.

[2] Bald, Robert, *View of the Coal Trade of Scotland*, 1808, p. 18, and Appendix,
p. 149.

licence for the sole use of the said engine being eight years complete next following and ensuing." The names of the "Committee appointed and authorized by the Proprietors of the Invention for raising Water by Fire" are:

John Meres of London Gentleman.
Thomas Beake of the City of Westminster Esquire.
Henry Robinson, Citizen and Mercer of London.
William Perkins, of the City of Westminster, Tallow Chandler.
Edward Wallin of London Gentleman.

These names, however, have not led to clues of any importance.

The second document, dated 1727, gives an account of the expenses incurred for materials for the engine, exclusive of the engine house, amounting to the very large sum of £1007. 11s. 4d. of which the cylinder alone accounted for nearly a fourth, viz. £250.

	£	s.	d.
Imprimis. To a cilinder 29 inches diameter with workmanship carried to London and all other charges and expenses	250	0	0

Other items of almost equal interest are worth quoting:

	£	s.	d.
To a pestion	17	10	0
To Elm pumps at London	53	4	6
To Two cast-mettle barrels 9 foot long and 9 inches diameter and with expenses after them . .	41	16	6
To the plumbers bill for lead and a lead tap for the boyler with sheet lead and lead pipes . . .	78	10	6
To the timber bought in Yorkshire for the engine with carriage by land and water and freight to Newcastle	82	16	0
To two brass buckets and two clacks 9 inches diameter, a brass regulator and injection cock and other cocks; sinking vouls, injection caps, snifting vouls and feeding vouls	35	5	0
To 44 cwt 1 qr 14 lb of chains, screw work and all other iron work about the engin except the hoops of the pumps at 5d *per* pound	103	10	0
To plates and revet-iron for making the boyler. .	75	10	0
To iron-hoops for the pumps with screw bolts and plates for ditto 18 cwt at 4d *per* pound . .	33	12	0

Then there were the expenses of the erectors:

	£	s.	d.
From July 12th to and with Dec 1 1726 being twenty weeks for Ben and Robin at £1 : 10s *per* week .	30 :	0 :	0

John Potter gave his receipt for the total on July 1, 1727.

At the end of the patent privilege, on Dec. 10, 1735, the sum of £240 is accepted "in full and compleat satisfaction of the whole obligement contained in the foregoing articles and annuity thereby to be paid to the Members of the Committee or their executors".

Abraham Potter, brother german (i.e. half-brother) of John, was appointed steward and factor of the colliery for the remainder of the lease; the engineers were to be paid £200 per annum to keep the engine going and to have half the profits of the colliery. If they did not succeed in making the engine draw water they were to be at liberty to take away all the material they had furnished and to be paid a reasonable allowance for their pains and charges. Such was the cost of draining a colliery in the year of our Lord, 1725.

The first point that deserves attention is that the duration of the licence is till 1733, the date of expiration of Savery's patent as prolonged by Act of Parliament, and the date of the expiration of James Smith's Scottish patent, affording conclusive evidence that Newcomen's invention was exploited under the former's patent.[1] It is this fact that has given rise to so much confusion between the two inventions; the engine is usually called the fire engine, the name given to Savery's apparatus, and Newcomen's invention is frequently called an improvement on Savery's.

The procedure in the erection of these engines is quite clear from this account and from other sources. An engine was built as a house is built to-day, i.e. materials were brought together on the spot from various localities, e.g. stone or brick from the neighbourhood, timber possibly from the Midlands, the cylinders and pump-work from London, etc., to be worked into position

---

[1] There is a slip in the date given for the expiration of the licence; it should be July 24, not June 24.

on the site by the different craftsmen, under the superintendence of the engineer. The various materials were paid for by the owner as were the wages of the workmen; an appropriate royalty was charged by the "proprietors" of the patent, and an annual sum for maintenance was paid to the engineer or his deputy. This method of erection and working persisted for over a century, as we shall see when we come to the Watt engine. Edmonstone, it may be remarked, was not the first engine in Scotland; it was preceded by at least two others, one of which is that mentioned in the licence.

About 1725 there appears to have been an increasing demand for the engine and several important engines were built. Some confirmation of this is afforded by the fact that a copper-plate engraving of the engine, by Sutton Nicholls in 1725, was published by John King in the Poultry in 1726, whether with the object of advertising the engine, or with the intention of giving agents of the "Proprietors" something to show to prospective customers, we do not know. In this engraving[1] there is a strong family resemblance to the Barney print. The valve gear is of the tappet type, and there is one improvement, whether due to Newcomen or not we do not know, that is, the boiler is fed with water from the eduction pipe instead of from the top of the piston. Desaguliers in 1744 thus describes it:[2]

It had been found of Benefit to feed the Boiler with warm Water coming from the Top of the Piston rather than cold Water, which would too much check the boiling, and cause more Fire to be needful. But after the Engine had been placed some Years, some Persons concerned about an Engine observing that the injected Water as it came out of the Eduction Pipe was scalding hot, when the Water coming from the Top of the Piston was just luke-warm, thought it would be of great Advantage to feed from the Eduction or injected Water, and accordingly did it...which gave a Stroke or two of Advantage to the Engine.

An engine was supplied in 1726 to York Buildings to replace the Savery engine there, which had proved so unsatisfactory; it

[1] Only three copies are known; one in the British Museum, one in the Science Museum, and one in the possession of the Naturforschende Gesellschaft, Danzig.   [2] *Experimental Philosophy*, II, 1744, p. 481.

resembled the engine in the above engraving, indeed the erection of the engine may have been one reason for publishing the print. Switzer also describes and illustrates an engine which differed, he says, in no essential point from that at York Buildings, which he calls a "Noble Engine".

As we have said the knowledge of the new power that had been born into the world had reached the Continent. We have already mentioned the engines in Hungary. It was natural from the advanced state of the industrial arts in France that the engine should be introduced there. The first of which we have a record was erected at Passy, then outside but now within the boundaries of Paris, to raise water from the Seine for supplying the city with water. The names associated with its construction are those of John May and John Meeres, and the date was 1726. The last named was almost certainly the Meres of the "Committee of Proprietors", so that on the assumption that Newcomen was working in collaboration with them it is barely possible that he might have visited France. The engine created much stir and was very fully described,[1] but it had no unusual features. The engine is stated to have raised about 25,000 muids (9,900,000 gallons or 1,587,500 cubic feet) of water in 24 hours but no height is given, so that we cannot calculate the horse-power.

An engine is stated to have been erected at a coal mine at Jemeppe-sur-Meuse, near Liège, prior to 1725, and to have been followed by others, but details are lacking. There was certainly a project to instal one at Toledo in Spain, for water supply of the city about the same time, but whether it was ever erected is doubtful. An engine was proposed for Sweden in 1725, but it was not till 1737 that the first engine there was actually erected. This was the engine of which Triewald, who has already been mentioned, was the engineer.

[1] *Machines approuvées par l'Académie des Sciences*, IV, 1726, p. 185. Cf. *The Engineer*, January 15, 1905, p. 129.

DEATH OF NEWCOMEN

Newcomen, while in London, presumably on business, died of a fever after a fortnight's illness, on August 5, 1729, at the house of a friend. Being by religious persuasion a Baptist, he was buried in Bunhill Fields, the Nonconformist burying ground, on August 8, but the exact spot is unknown because the ground has since been buried over again. Only one document in his handwriting and a signature or two have come down to us. We have no portrait of him, and of his private life we know practically nothing. Unlike Savery, he did not become a Fellow of the Royal Society, and he did not write a book. He had lived to see his engine in use throughout England, Scotland and Wales, and in Hungary, France, Belgium and possibly Germany and Spain, but whether he derived any pecuniary advantage from his labours is open to grave doubt. It is difficult to resist the conclusion that he had been thrust aside by the "Proprietors" aforesaid, who had taken the lion's share of any profits that accrued from the erection of engines on his principle.

The "Mr. Calley" who was concerned with the erection of the engine at Austhorpe died there in 1717, and a John Calley died at The Hague in 1725, but what relation they were to one another, or to Newcomen's partner, we do not know. About the latter, again, we know nothing personally, nor what actual share, if any, he had in the invention of the atmospheric engine.

Thus the men directly connected with the invention and introduction of the engine had died, the "Committee of Proprietors" had dissolved at the expiration of the patent, and it was now open to anyone to make the engine. The work of doing so fell into the hands of practical men, who could handle the hammer but not the pen, while the men of science of the time neglected the engine, either blind to the fact that vast control was thereby being established over the forces of nature, or disdainful of the mechanic arts as being beneath their notice.

As an appropriate ending to this chapter, we make no apology for introducing the story of the steam engine up to this point, as told in verse, in the form of an "Ænigma" which appeared in the *Ladies' Diary* of 1725,[1] and therefore presumably written by the editor, who as we have said was Henry Beighton. In the issue of the *Diary* for the year following that in which the Ænigma appeared,[2] the following explanatory note appears:

The Prize Ænigma is a Description of the Invention and Progress of the Engine for raising Water out of Mines by the Force of Fire. It was first used by *Herbert* Marquiss of *Worcester*, about the year 1644, and published in his *Century of Inventions* anno 1661. In 1698, Capt. *Tho. Savery* got a Patent for 14 Years, and an Act of Parliament for 21 Years longer for that Invention. In the year 1712 Mr. *Newcomen* by applying the Weight of the Atmosphere instead of the Elasticity of the Steam, brought it to the Perfection wherewith it is now used.

The "Prize Ænigma" runs thus:

ÆNIGMA.

I Sprung, like Pallas, from a fruitful Brain,
  About the Time of *CHARLES* the *Second's* Reign.
My Father had a num'rous Progeny,
And therefore took but little Care of me:
An Hundred Children iſſu'd from his Pate;
The Number of my Birth was Sixty Eight.
My Body ſcarcely fram'd, he form'd my Soul,
Such as might pleaſe the Wiſe, but not the Dull:
Yet ſundry Pictures of my Face he drew;
As many of his other of his Children too:
Theſe Pictures lay, whilſt none my Worth did know,
In *Paul's Church-Yard*, and *Pater-noſter Row.*
My Father Dead, my ſelf but few did ſee,
Until a Warlike Man adopted me;
Deſtroy'd what Records might diſcloſe my Birth,
Said *He begot me*, and proclaim'd my Worth.
Begetting me he call'd a Chance— A Task
Eaſie to him, aſſiſted by a Flask.
  He then to me ſtrange Education gave,
Scorch'd me with Heat, and cool'd me with a Wave:
More Work expected from my ſingle Force,
Than ever was perform'd by *Man or Horſe.*
To mend my Shape, he oft deform'd it more;
Which ſometimes made me Burſt, and Fret, and Roar:
Then from my Eyes, ſuch Vapours iſſu'd forth
As *Comets* yield, or Twilights of the North:

---

[1] Pp. 18, 19.   [2] *Loc. cit.* p. 10.

And like thofe Lights, the Vulgar I furprize;
Not thofe that know my Nature, or the Wife.
   My Heart has *Ventricles*, and twice Three *Valves*;
Tho' but One *Ventricle*, when made by Halves.
My *Vena Cava*, from my further Ends
Sucks in, what upward my great *Artery* fends.
The *Ventricles* receive my pallid Blood,
Alternate; and alternate yield the Flood:
By VULCAN's Art my ample Belly's made;   ⎫
My Belly gives the Chyle with which I'm fed; ⎬
From NEPTUNE brought, prepar'd by VULCAN's Aid. ⎭
My Father (I mean He who claim'd my Birth)
My Dwelling fix'd in Caverns of the Earth;
And there, he faid, I fhou'd in Strength excel;
But there, alas! I was but feldom Well.
Torrents he bad me ftop: —— I wanted Breath;
And Nature ftrain'd too much, will haften Death.
In this fad State, to languifh I begin,
Until a Doctor fage, new coming in,
Condemn'd the Methods that were us'd before;
And faid, — That I in Caves fhou'd dwell no more:
Then I fhou'd dwell in free and open Air,
And gain new Vigour from the Atmofphere:
An Houfe for me he built —— Did Orders give,
I fhou'd no Weight above my Strength receive;
And that I fhou'd, for Breath, and Health to guard,
Look out of Windows when I labour'd hard.
   Thefe gentle Means my Shape have alter'd quite;
I'm now encreas'd in Strength, and Bulk, and Height;
I can now raife my Hand above my Head;
And now, at laft, I by my felf am fed.
   On mighty Arms, alternately I bear   ⎫
Prodigious Weights of Water and of Air;   ⎬
And yet you'll ftop my Motion with a Hair.   ⎭
   He that can fine me, fhou'd rewarded be, ⎫
By having, from my Mafters, Liberty,   ⎬
Whene'er he pleafes, to make ufe of me.   ⎭

From what has gone before, the allusions will be readily understood, "My Father" is the Marquis of Worcester; "a Warlike Man", Capt. Savery; the *"Ventricle"*, the receiver; the *Vena Cava*, the suction pipe; the *"Artery"*, the force pipe or rising main; the "Blood", the water; the "Chyle", the steam; "a Doctor fage", Newcomen, and "new coming in", a pun on his name.

CHAPTER IV

THE ATMOSPHERIC ENGINE IN THE PERIOD BETWEEN NEWCOMEN AND WATT

Rebirth of iron smelting in England—Increase in size of engines—Boiler capacity—Cost of erection and operation—Smeaton's experiments—His improved engine—His water coal-gin—The atmospheric engine modified for winding from mines.

FOR sixty years, that is to say from 1712 onwards, the atmospheric engine remained the only efficient means of draining mines or supplying towns with water, but as regards improvements in the engine, apart from those introduced by Newcomen himself, this period might almost be passed over in silence as a barren one, were it not that it was a period of extraordinary technical advance in nearly every industry in this country, an advance in which the steam engine, in the limited role that it played so far, was of importance because by its means coal was cheaply won, and such fuel was required for nearly every industry. This advance was particularly marked in the case of iron manufacture, a matter, in its turn, of the greatest importance to engine construction. What amounted to a rebirth of iron smelting in England took place in the first half of the eighteenth century. Originating with the substitution of mineral fuel for charcoal for smelting, made by Abraham Darby round about 1713, at Coalbrookdale, Shropshire, the coke pig iron so produced was used, unlike the charcoal pig iron, not for making wrought iron, but for casting bellied pots, and this led on to casting hollow iron articles generally; in other words, to iron-founding on a large scale, so that about 1730, cast-iron engine cylinders and pipes began to be cheap and readily procurable, reducing thereby the initial cost of installing engines. Instead of paying £250 for a brass cylinder as was necessary in the Edmonstone engine already referred to, by 1740 cast-iron cylinders could be had from Coalbrookdale at £20 to £30 apiece. Not only so, but iron plates hammered under a tilt hammer became a commercial article and boilers were made of the material—Stanier Parrot of Coventry is credited with the innovation. Thus iron, both cast and wrought, came more and more into use. Brass cylinders, copper boilers and wooden pump-trees became things of the past, not, however, without the usual opposition from the diehards. Among them, for instance, was Desaguliers, who says:[1]

SOME People make use of cast Iron Cylinders for their Fire-Engines; but I would advise nobody to have them, because tho' there are Workmen that can bore them very smooth, yet none of them can be cast less than an Inch thick, and therefore they can neither be heated nor cool'd so soon as others, which will make a Stroke or two in a Minute Difference, whereby an eighth or a tenth less Water will be raised. A Brass Cylinder of the largest Size has been cast under 1/3 of an Inch in Thickness; and at long run the Advantage of heating and cooling quick will recompense the Difference in the first Expence; especially when we consider the intrinsick Value of the Brass.

Desaguliers was right; it was obvious that heat and consequently coal could be saved if the bulk of metal to be heated up and cooled down at every stroke was kept small, and therefore a cylinder $\frac{1}{3}$ in. thick, as he mentions, was much to be preferred to one of 1 in. thickness. It was original capital cost, however, that decided the issue and cast iron won the day. Coal, generally speaking, was not a large item in cost, as quantities of small coal, practically unsaleable but good enough for boiler firing, were dumped at the pit-heads where the engines were mainly to be found. In metalliferous mining districts such as Cornwall, on the contrary, the cost of coal which had to be imported was a formidable item. Consequently, the engine did not make much headway there until 1741, when after eleven years of agitation an Act of Parliament was passed waiving the duty of 5s. 5d. per chaldron charged on coals when the latter were used for pumping in mines. Even with this concession, the cost of the coal was almost prohibitive, and it was in Cornwall that the search for cheaper pumping power plant became most insistent. However, needs must when the devil drives, and by 1758 engines up to 70 in. diameter cylinders were erected, and by 1778 about sixty had been supplied.

[1] *Experimental Philosophy*, II, 1744, p. 536.

Probably progress in adopting the engine was most rapid in the coalfields of Durham and Newcastle. This was a direct consequence, there can scarcely be a doubt, of the disastrous flooding of the Tyne coal basin towards the end of the seventeenth century, when many collieries were drowned out and, unless such means as the new engine had become available, must have remained so. A rebirth of the Tyne coal industry owing to this and the greater depth to which mining could be prosecuted now took place.

It is possible that there were still some coal owners and lessees who were not yet convinced of the enormous saving in cost by employing the new engine, and it is for them probably that the following estimate was drawn up:

Estimate of the difference of the Expence in Drawing Water by Fire Engine & Drawing it by Horses, made Dec. 11, 1752.[1]

The document is too long to reproduce in full but it may be summarized by taking the comparative totals thus:

40 *Fathom Shaft.*

Fire Engine.	*Horses.*
Working Barrells 12 Ins. diam. Stroke of Engine 5 ft. 8 Strokes in a minute.	2 Horses to work 3 Hours for a shift will draw 35 Tubbs. Each Tubb containing 80 Gallons.
In 24 Hours will need 6 hours for drawing and putting in buckets & clacks, repairs to shaft, cleaning and repairing Boiler etc. Stops, accidents etc., leaving 18 hours to draw water in each 24 hrs.	
In 24 hours will draw 250,560 *galls.*	In 24 hours will draw 67,200 *galls.*
Suppose Annual Expence for a Fire Eng. to be £365 (including interest on cost of Erecting, etc.).	8 shifts of horses 2 in a shift will deserve, including Driver, 3 sh. per shift.
Cost per 24 hrs will be 20 *sh.*	Cost per 24 hrs will be 24 *sh.* & does not draw $\frac{1}{3}$ the water.

[1] North of England Institute of Mining Engineers, MS. Mining Collection. Cf. *Trans. Newcomen Soc.* XVII, p. 153.

It will be noted that this is an estimate of running cost, so that it is incomplete; the interest on the capital cost of the plant and animals would alter the figures, but enough is set out to show that the saving was of the order of 30–40 per cent.

A list of engines existing in the North of England in 1769 taken from the books of William Brown of Throckley, a well-known "viewer", as the colliery engineer was called, shows that there were then ninety-nine in existence, fifty-seven of which were at work and the largest of them had a diameter of 75 in. This list was probably prepared at the instance of John Smeaton, mentioned below.

In France we ought to mention an engine that has some claim to distinction, as it was chosen by the celebrated architect and engineer, Bernard Forest de Bélidor (1697–1761), for illustration and description in his well-known work, *Architecture Hydraulique*, 1739. This was the engine at a colliery at Fresnes near Condé sur l'Escaut, in what is now the Département du Nord, France, erected by English engineers some time before the publication of that work. Previous to the installation of the engine, the colliery was kept unwatered by animal power; fifty horses and twenty men in all, working in shifts throughout the 24 hours, had been necessary, whereas with the engine and two men, a week's water could be raised in 48 hours. The engine, which had a cylinder 30 in. diameter and 9 ft. long, drew from a depth of 90 ft. 410 cubic ft. per hour at the rate of 15 strokes per min. Bélidor attributes the invention, as was to be expected, to Savery, but he does say that, in his correspondence with the Royal Society, he found that Newcomen was credited with having brought the engine to perfection. This shows by the way that Newcomen had come to be recognized in his own country by this time as the inventor. Bélidor is loud in his praise of the engine; his words are so apt and expressed in such quaint conceit that they are worth quoting:

[Translation.] It must be confessed that these are the most marvellous of all engines, and that there is not one of which the mechanism has more agreement with that of animals. Heat is the principle of its movement; in its various pipes it creates a circulation like that of the blood in the veins, having valves which open and shut opportunely; it feeds itself; it discharges itself at regulated times and draws from its own work all that it needs to subsist.

While on the subject of encomiums on the engine, we feel we must quote the following, accompanying an engraving[1] and description of it, because incidentally the correspondent mentions a number of engines in London and the North, not hitherto cited, showing how widespread the use of the engine had become by this time:

> ...*I have sent you inclosed a* Draught *and* Description *of an Engine invented for this Purpose* (i.e. for draining of mines) *and, though there are many other Sorts, I have rather selected this particular* Engine, *because it is the most admirable, curious and compounded Machine amongst all those Inventions, which have been owing to modern Philosophy, and affords the greatest Advantages to mankind; as could be exemplified from the* Water Works *near* Chelsea, *on the* West *of this great City, and again by those lately erected near* Stratford *in* Essex *on the* East *of* London, *which are able to supply the adjacent Country several Miles in Circumference with the necessary Provision of good and wholesome Water, at a moderate Charge, which before was wanting both for household Service, and in the Danger and Loss by Fires. To this I could add the Impossibility of working several* Colliaries *without its Assistance, as the Proprietors of* Elsick, Heaton, Biker, &c. *near* Newcastle on Tyne, *can bear me witness. This Engine also is improveable for many other great and valuable uses, as the Reader will be able to judge, when he has well considered what follows.*

As the size of engines increased, the single boiler, situated immediately underneath the cylinder, which had been Newcomen's practice, was found to be inadequate, for it is to be remembered that the engine was a great "steam-eater". Hence, two or more boilers symmetrically placed relative to the cylinder began to be used. Gabriel Jars (1732–69), the well-known French metallurgist and traveller, who visited Newcastle-upon-Tyne in 1765, describes the engine at Walker Colliery—the largest in the area at that time—built by William Brown, then the best-known viewer or colliery engineer in the district. It

---

[1] *Universal Magazine*, I, 1747, p. 162. The engraving adds nothing to our knowledge and has not been reproduced.

had no less than four boilers, three of which were in service, and the fourth spare—the first instance known to us of the provision of spare plant to ensure continuous working. Jars notes several other points of interest: the boilers were made wholly of wrought iron with lead tops, except in the case of the boiler underneath the cylinder, which had to have a copper top because to it had to be attached the steam admission pipe and valve; the piston was packed with hemp rope; three jets were used for condensation in the cylinder on account of its size. From another source we learn that the diameter of the cylinder was 74 in. and its length 10½ ft., that it weighed 6½ tons and came from Coalbrookdale. We marvel at the skill with which the engineers of those days handled such great weights with their primitive lifting and transporting appliances, particularly when transhipping from a vessel to the quay. We learn that "The bore [i.e. of the cylinder] was perfectly round and well polished". In this connection it ought to be repeated that when Newcomen started engine-building, the means for boring cylinders were practically non-existent. It is true that guns, both bronze and cast iron of small calibre, and pump barrels up to 7 in. diameter were being bored, but when it came to 19 in. diameter, the size of the first engine cylinder, the task was too heavy, and the first cylinders are said to have been rubbed smooth on the inside with abrasives. By the time of which we speak, boring mills of greater size and power were available, with the result that cylinders could be supplied as stated; small wonder, however, that they were so expensive.

It will have been gathered already from the accounts of the Edmonstone engine what an enormous expense was entailed by the installation of an atmospheric engine. Obviously, such an expense could only be justified where, as in the North of England, the expectation of life of a colliery was such as to cover the cost; alternatively, there was the possibility of removing the engine to a new pit, and we find this done quite frequently in practice. In the MS. Mining Collection of the North of England Institute of Mining Engineers at Newcastle-upon-Tyne, from which we have quoted already, there are many estimates of the cost of erecting fire engines. One such estimate of date 1733 is as follows:

An Estimate of Charge of a Fire Engine

The Bore of Pumps that will Discharge 400 hhds. in One Hour must be 13 Inches Diameter also the Diameter of the Cylinder 33 Inches to run 12 Strokes p. minuet ye Strokes 6 Foot Long.

To 22 fathom of Pumps 12 Inches Diameter . .	£30 : 00 : 00
To Boreing joynting pluging Hooping & Screw bolts to fasten the Plugs in the pumps . .	24 : 00 : 00
For Two Brass Barrels for the Bages to Work in 13 Inches Diameter	90 : 00 : 00
To Two Buckets and Two Claks	6 : 00 : 00
To Iron to hoop the pumps	37 : 00 : 00
To 40 Fathom Spears & joynts belonging to them	19 : 00 : 00
To the Cistern & Slipers to bare the piller of Pumps	5 : 00 : 00
To Smith work	24 : 00 : 00
To the Regulater beam with ye Arches to it . .	24 : 00 : 00
To the Gudgeon that Carries the Beam, Stirrups, barrs and screw plates	3 : 15 : 00
To Two Brasses that the Gudgeon runs in . .	6 : 00 : 00
To 3 Chains, Catch pins, & 4 Plates Belonging to the Regulater Beam	26 : 00 : 00
To the Engine House	120 : 00 : 00
To 33 Inches Cylinder	150 : 00 : 00
To 2 Cylinder Beams, 2 plates & 7 screw bolts .	27 : 00 : 00
To the Piston	3 : 00 : 00
To the Boyler	80 : 00 : 00
To Leading the Boyler Top	32 : 00 : 00
To the Regulater that stands at the Top of the Boyler	9 : 00 : 00
To the Sinking Pipes	12 : 00 : 00
To the Other Pipes About The Cylinder . . .	8 : 00 : 00
To the small the F and Y & 2 Small Brasses ye Y runs in	2 : 06 : 00
To Cocks About The Cistern	3 : 15 : 00
To small Valves	0 : 10 : 00
To Bars for the Fire place	12 : 00 : 00
To Bricks, Lime and Workmanship	13 : 00 : 00
To a Crab Rope	8 : 00 : 00
To 2 Shear Legs & a Block	10 : 00 : 00
To a New Crab	2 : 10 : 00
To Soder	15 : 00 : 00
To Jack head barrells & Pumps	35 : 00 : 00
To Deals to Make ye Great Cistern & Workmanship	12 : 00 : 00
	849 : 16 : 00

A dozen years later the size of engines had increased and the cost was about doubled. From a detailed estimate of the "Expence of Walbottle Fire Engine" of about 1745, we extract the following: The cost of the engine and engine house was upwards of £1050, of which the engine house cost £500, and the cylinder with piston and bottom alone amounted to £208; the rest of the costs—the largest item in which was the cost of iron pipes, £335—amounted to £640, so that the total cost was £1691. 15s. 6d.—a very large sum with money at its then value.[1] The annual cost for coals and attendance was about £400. That the cost was justified is realized when the expense of pumping by fire engines was compared with that of horse whims which they superseded (see p. 56), in the estimate already cited.

We have already mentioned that John Smeaton paid much attention to the atmospheric engine. In the course of his professional work he was called in to advise upon and to design many pumping engines. He was dissatisfied with both the design and the performances of the engines that came under his observation. Now Smeaton was a man who could not rest until he had made himself master of a subject. Accordingly, he set up in the workshop attached to his house at Austhorpe, near Leeds, in 1765, an experimental model engine in which he broke away from established practice in that, in the first place, the engine had no beam, its place being supplied by an oscillating wheel; in the second place, the model was portable; and in the third place, the boiler had an internal flue.[2] With this model, having a cylinder of 10 in. diameter and 38 in. stroke, he carried out experiments and seems to have arrived at the conclusion that the cylinder should be small in diameter in relation to its length. In an engine which he constructed for the New River Co. at New River Head, Islington, in 1767, he put his ideas into practice but found that they did not answer his expectations. He accordingly went to work systematically, with the aid of William Brown of Throckley, and obtained from him the list already mentioned of about 100 engines, mainly in the North of

---

[1] *Trans. Newcomen Soc.* XVII, pp. 136, 153.

[2] Smeaton's *Reports*, I, p. 223.

England in 1769. Smeaton obtained particulars of the performance of fifteen of these, from 20 in. to 75 in. diameter of cylinder, and estimated that averaged out they raised 5·59 millions of pounds of water 1 ft. high by the consumption of 1 bushel of coal (84 lb.); the average pressure on the piston was only 6·72 lb. per square in.[1] He also collected particulars of eighteen large engines in Cornwall. In the winter of 1769 he conducted a series of experiments with the model at Austhorpe, and compiled in 1772 a table of proportions of the parts of engines with cylinders up to 72 in. diameter.[2] Based upon these data, he designed and superintended the construction of quite a number of engines. The first of these was at Long Benton Colliery, Northumberland, 1772, cylinder 52 in. diameter (see Fig. 9), and the most famous was that at Chacewater, Cornwall, 1775, cylinder 72 in. diameter. Another almost as famous was that for emptying the docks at Kronstadt, Russia, 1775, cylinder 66 in. diameter.

Smeaton, by improving the design and the execution of the parts, especially the boring of the cylinder for which he built an improved boring mill at Carron Iron Works, succeeded, in the Long Benton engine, in attaining a duty of 9,450,000 lb. raised 1 ft. high by the consumption of 1 bushel of coal (84 lb.), and this he looked upon as his best, and also his standard, performance. In other words, by his improvements he doubled the performance of the atmospheric engine, and it is not too much to say that what he attained was the highest performance that the engine was capable of in practice. Smeaton is worthy therefore of being included among the pioneers of the steam engine. The portrait which we reproduce (see frontispiece) is that attributed to J. Richardson in the possession of the Royal Society; the date of the portrait is not known.

It will be observed that thus far the engine was incapable of being used for anything except raising water, i.e. its motion was purely up-and-down or reciprocating. It was inevitable that many minds should be attracted to the problem of how to change

[1] Farey, *Steam Engine*, Fig. 2.
[2] *Ibid.* p. 183.

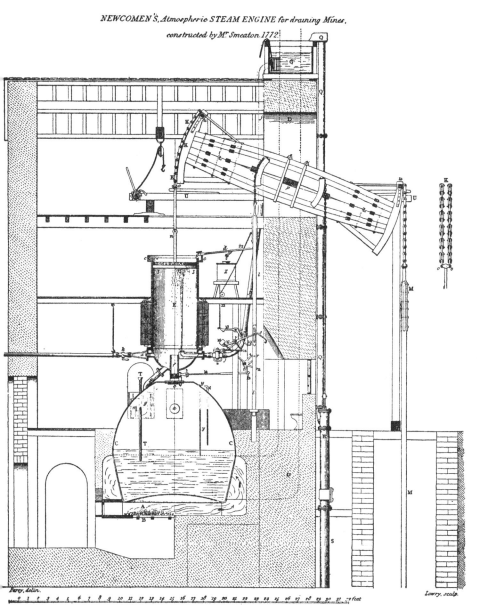

Fig. 9. Smeaton's atmospheric engine at Long Benton Colliery,
Northumberland, 1772.

From Farey's *Steam Engine*, 1827.

this reciprocating motion into a rotative one; in other words, how the engine could be made to turn a shaft. When the giant power of the engine was exhibited side by side with the puny power of the horse whim or of the water gin, which was all that was available for winding coal or minerals from the shafts of a pit or mine, it will easily be understood that persons would attempt to get the engine to turn the winding drum instead of the horse or water doing the work. We may mention one such attempt, that of Joseph Oxley, who patented in 1763 the application of a ratchet mechanism whereby the shaft was turned in one direction in the outdoor stroke, and the ratchet overran the shaft in the indoor stroke of the engine. This he patented in 1763, and applied it to raising coals at Hartley Colliery, Seaton Delaval, Northumberland, but it was extremely clumsy and got out of order continually, so that eventually the engine was altered to the ordinary construction, and made to pump water on to a waterwheel which itself did the pumping. This indirect method was used somewhat widely, at any rate for a time, in the North of England. Smeaton made a particularly neat design in 1777 for the Prosperous Pit, Long Benton Colliery. Water was pumped into an elevated cistern and was directed on to an overshot waterwheel, one half of the width of which had a right-handed set of buckets, and the other half-width had a left-handed set. By directing the water first on to one set of buckets and then on to the other, motion could be given in either direction. This machine was known as a "water coal-gin". Indeed, pumping water on to a waterwheel was the only known practical way of obtaining rotative motion from an engine for many a long year, and was practised even after the introduction of the rotative engine by Watt, to be described in the next chapter, because of the extreme uniformity of turning movement thereby assured—a matter of the utmost importance in such manufactures as cotton-spinning.

Towards the close of the eighteenth century, the atmospheric engine was actually made rotative by using a connecting rod, crankshaft and flywheel; the difficulty presented by the flexible chain connection of the beam to the piston was overcome by the expedient of loading the connecting rod, or else that end of the beam, with a weight approximately equal to half the load on the piston. The effect of this was to equalize the load on both the up and down strokes, but the expedient was a clumsy one. Such engines, on account of their cheapness, were made in large numbers for winding from shallow pits. They were known in the Midlands as "whimseys", and they were widely used even as late as 1820. Indeed, a very few survived into the present century—one such, fitted however with a parallel motion, that the author has seen at work, was at Farme Colliery, Rutherglen, near Glasgow. It was built in 1810 and remained in use till 1915, when it was presented to that city, and re-erected in Kelvingrove Park. The valves were operated by hand, and smartness on the part of the engineman was required to stop the engine at the right point where rotation could be resumed, as otherwise the engine would stall.[1] This was done by the aid of a band brake on the winding drum, and this brake also served to let the cage down into the pit.

It will be realized that by the third quarter of the eighteenth century the atmospheric engine had got to the end of its tether and that the industrial world was ready for a big advance. This is the subject-matter of the next chapter.

[1] *Proc. Inst. Mech. Eng.* 1903, p. 681.

CHAPTER V

WATT AND HIS SEPARATE CONDENSER ENGINE

Watt's career—Repairs atmospheric engine model—Stumbles on latent heat—Invents and patents separate condenser—Roebuck helps with experimental engine—Boulton takes over patent—Watt removes to Birmingham—Extension of patent—Partnership with Boulton—Engines in Cornwall—Premium—Compound engine—Rotative engine—Substitutes for crank—Governor—Expansive working—Indicator—Infringements—Partnership expires—Watt summed up—Economic advance.

THE atmospheric or vacuum engine, no longer a marvel to the unlettered, neglected by men of science, and its practical construction relegated to the hands of millwrights, had gone on the even tenor of its way for years, unruffled by any event until 1765, when an invention was made that was destined to change its history. The invention was that of the separate condenser, the greatest single improvement ever made in the engine. The inventor was an obscure mathematical instrument-maker in Glasgow, named James Watt (1736–1819); what he did, to put it in a nutshell, was to effect the condensation of the steam, not in the cylinder itself, but in a vessel separate from it, thereby conserving most of the heat that had hitherto been thrown away by the alternate heating and cooling of the cylinder. Thus at one bound he effected a saving of almost 75 per cent in fuel. He made other inventions later, as we shall see, but everyone of them was rooted in this condenser. A new era was thus inaugurated amounting almost to a rebirth of the engine, but by keeping in mind firmly the facts that have been recited previously we shall avoid the common error of crediting Watt with the entire invention of the steam engine, an error that is excusable in view of the outstanding character of the numerous improvements that Watt actually did introduce into the vacuum engine.

It is interesting to pause for a moment and reflect upon the differentiation of function of working parts of the steam engine that had taken place (see Fig. 10). Papin had a piston in a cylinder in which he boiled his water, afterwards condensing the steam slowly in the same vessel. Newcomen generated his steam in a vessel separate from the cylinder, but still condensed the steam rapidly in the cylinder itself. Watt made use of a boiler and a cylinder like Newcomen, but condensed the steam rapidly in an entirely separate vessel. One sees here almost a biological analogy.

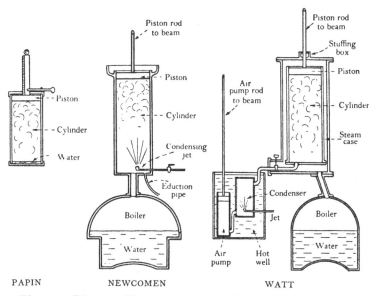

Fig. 10. Diagram illustrating progressive differentiation of function in the steam engine.

To make it clear how it came about that Watt was drawn to the study of the steam engine it is necessary to state the leading facts that led up to that stage, although they are already well known.[1] He was a delicate, studious lad, and when the time came for him to earn his living, he expressed a desire to take up mathematical instrument-making, as he had already had some leaning thereto; the difficulty was, however, that the trade was not then carried on in his native town of Greenock, or even in the nearest city, that of Glasgow. Hence he was advised by Dr Robert Dick, Professor of Natural Philosophy in the College

[1] Cf. Dickinson, *James Watt, Craftsman and Engineer*, 1935, p. 21.

of Glasgow, to whom Watt had obtained an introduction through his kinsman Professor George Muirhead, to go to London to obtain instruction in his chosen craft. With his father's aid he managed to do this and spent a strenuous year in the metropolis, returning to Greenock in August 1756 provided with a kit of tools, proud to consider himself a finished craftsman. In October of that year he was taken on at the College of Glasgow by Dr Dick to clean some astronomical instruments that had come from the West Indies, and in consequence he was allowed to set up a shop in part of the College buildings, opening on the High Street where the College at that time was situated. Watt found the atmosphere of learning congenial and made many friends among the professors and students. We can mention only Dr Joseph Black, Professor of Chemistry, and John Robison, who had just graduated Master of Arts. Dr John Anderson, who succeeded Dr Dick as Professor of Natural Philosophy in 1757, gave Watt the job of repairing a model of an atmospheric engine which belonged to the Natural Philosophy class of the College. The model "had been in such a situation that it did not answer the end for which it was made". It had, in fact, been up in London at a well-known mathematical instrument-maker named Jonathan Sisson (c. 1690–1760), but he had not effected any improvement. Not so Watt—he made it work and in doing so made himself master of its principle of action. He was struck with its consumption of steam, which was so great that with the available boiler capacity no more than a few strokes of the engine could be obtained before the model was obliged to stop. He realized that this was due to the waste of steam, caused by the alternate heating of the cylinder to boiling point and its cooling to a room temperature at each stroke. He made an experimental boiler and found *inter alia* that the quantity of steam used by the model at each stroke was several times the volume of the cylinder. Watt then undertook some fundamental experiments, about which we shall have more to say in a later chapter. For example, he passed steam into water at the ordinary temperature until the water became boiling hot and found that it had gained about one-sixth in volume. Puzzled by this apparent disappearance of heat, Watt at this juncture learnt from Black of his discovery shortly before of the phenomenon of latent heat, i.e. that bodies in changing their physical state, say, from ice to water or from water to steam, give out or absorb heat without it becoming sensible to the thermometer, or, in other words, it lies latent. One Sunday in May 1765, after months of racking thought, while out for a walk on Glasgow Green, the idea flashed into his mind "*that as steam was an elastic body it would rush into a vacuum, and if a communication was made between the cylinder and an exhausted vessel, it* (the steam) *would rush into it and might be there condensed, without cooling the cylinder*".[1] Thus was begotten the separate condenser, an organ that, vital as it was to the economy of the vacuum engine, is even more so to-day to that of the steam turbine. As this momentous day was a Sunday, Watt had to wait, such was the Sabbath convention of the time, till the following day to make experiments. "In a few days he had a model at work with an inverted cylinder which answered his expectation." There is a model of Watt's in existence, now in the Science Museum, South Kensington, which, if not the identical model, must have been made about this time, for it bears the marks of hasty improvization and we cannot be far wrong in venerating it as the original model with which Watt demonstrated the soundness of his invention (see Fig. 11); whether by accident or design the connection has been soldered up wrongly, and a conjectural restoration is shown in Fig. 12. *A* is the cylinder, *B* the steam inlet and steam case, *C* the cylinder cover, *D* the piston, *F* the condenser, *H* the air-pump barrel and *G* the snifting valve.

Obviously we have here only the bare bones of an engine. Everyone who has had to deal with an invention knows, even though the public do not realize it, that it is a far cry from the idea, or even a model embodying the idea, to its realization as a working machine on a full-size scale. We have to remember that Watt knew practically nothing about the atmospheric engine; there is even a doubt whether he had at this time seen one; all he knew was what he had gleaned from experimenting with the model itself, and from books like Desaguliers's *Natural Philosophy*, or Bélidor's *Architecture Hydraulique*. Watt's ideas were

[1] Narrative of Robert Hart from Watt's own lips, *c.* 1813.

entirely original, as the model shows, e.g. his engine was to be inverted and direct-acting, and he spent some time in trying to realize this arrangement in practice. Gradually, however,

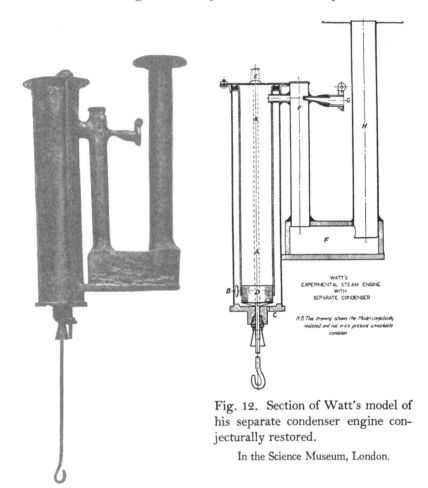

Fig. 12. Section of Watt's model of his separate condenser engine conjecturally restored.

In the Science Museum, London.

Fig. 11. Watt's original model of his separate condenser engine, 1765.

In the Science Museum, London.

working along the line of least resistance, and obtaining some practical knowledge of the construction of an actual engine, he arrived at the idea of applying the condenser to the existing beam engine. Through his college friend, Dr Black, Watt got into touch with that industrial pioneer and captain of industry, John Roebuck, of Birmingham and of Carron Iron Works, who was anxious to get improved means for unwatering the coal mines he was opening out at Bo'ness, near to Grangemouth. Watt, acting on the advice of his friend Black, sought protection for his invention by patent—the ever memorable one entitled "a new method of lessening the consumption of steam and fuel in fire engines", which was granted on January 9, 1769 (No. 913). Roebuck, in consideration of the funds he had already supplied to Watt, came in as a partner to it to the extent of two-thirds. An experimental engine, quite a small affair, with a cylinder 18 in. diameter by 5 ft. stroke, was built in an outhouse close to Dr Roebuck's residence, Kinneil House, but the scheme was not prosecuted vigorously, partly because of Watt's engagements as a civil engineer, by which he earned his bread and butter, and partly because Dr Roebuck was becoming financially embarrassed with his over-numerous commitments, and he could not spare the necessary funds for the exploitation of the invention. Watt had, however, through his friends, made the acquaintance of the "princely" Matthew Boulton, the industrial leader of the Midlands, whose Manufactory at Soho, near Birmingham, was celebrated all over Europe for its products.[1] Watt had visited Boulton in 1769, when on his way to London to take out his patent, and had been charmed with Boulton, and all his works. The latter had been privy to the patent negotiations, and had dismissed as impractical Roebuck's proposal to take up the manufacture of the engine for a limited area in the Midlands, although prepared to make for all the world. In a sympathetic and statesmanlike letter dated February 7, 1769, he showed not only clear insight into the particular circumstances of the case, but also enunciated incidentally the principle of mass production with which we are so familiar to-day. The letter is so revealing that in spite of the fact that it has been quoted repeatedly, we give it in full:

...the plan [i.e. Roebuck's three Midland counties' licence] proposed to me is so very different from that which I had conceived at

[1] Cf. Dickinson, *Matthew Boulton*, 1936, for further details.

the time I talked with you upon the subject that I cannot think it is a proper one for me to meddle with as I do not intend turning engineer. I was excited by two motives to offer you my assistance which were love of you and love of a money-getting ingenious project. I presumed that your engine would require money, very accurate workmanship and extensive correspondence to make it turn out to the best advantage, and that the best means of keeping up the reputation and doing the invention justice would be to keep the executive part out of the hands of the multitude of empirical engineers, who from ignorance, want of experience and want of necessary convenience, would be very liable to produce bad and inaccurate workmanship; all of which deficiencies would affect the reputation of the invention. To remedy which and produce the most profit, my idea was to settle a manufactory near to my own by the side of our canal where I would erect all the conveniences necessary for the completion of engines, and from which manufactory we would serve all the world with engines of all sizes. By these means and your assistance we could engage and instruct some excellent workmen (with more excellent tools than would be worth any man's while to procure for one single engine), could execute the invention 20 per cent. cheaper than it would be otherwise executed, and with as great a difference of accuracy as there is between the blacksmith and the mathematical instrument maker. It would not be worth my while to make for three counties only, but I find it very well worth my while to make for all the world.

Dr Roebuck was the obstacle, but when his long staved-off bankruptcy came about in 1773, Boulton was able to take over from the receivers of the estate, in discharge of a debt that Roebuck owed to him, the two-thirds' share that the latter possessed in the patent.

Of the total period of nine years since Watt had specified his improvements, only a matter of three years had been devoted to active prosecution of the invention; in this interval the engine had been assimilated to the atmospheric beam type, but it had not by any means attained complete success. Watt migrated with his family, in 1774, to Birmingham, bringing with him the experimental engine from Kinneil House; a new era now dawned for him, for he was now able to give it undivided attention and carry out intensive experiments with it, with such success that he was able to write to his father in 1774: "the fire engine I have invented is now going and answers much better than any other that has yet been made."

Now at last the way seemed clear, but a further difficulty presented itself—the patent had only eight more years to run, and Boulton with his business acumen realized that the patent would expire before the profit-making stage could be reached. The obvious thing to do was to try and get an extension of the patent. A petition to the House of Commons to bring in a bill to extend the patent for twenty-five years was granted; the bill was passed through all its stages, not, however, without violent opposition, and received the royal assent on May 22, 1775. The famous partnership of Boulton with Watt coterminous with this extension was entered upon on June 1 following.

The time was more than ripe for the introduction of Watt's improved engine. The change in the manner of production of commodities from handicraft methods to those of the factory, frequently called the industrial revolution, had begun and was creating a demand for power in large units. This demand was being met by harnessing of water power, and this led industry into the remoter districts where such power could be made available. Power brought to the door, so to speak, and made available in desired spots, unaffected by climatic considerations and capable of indefinite expansion, such as was that of the steam engine, was an advance of the greatest importance to industry. The knowledge of Watt's engine had got abroad, and many persons, Boulton himself among them, were waiting anxiously for the result.

Watt was still carrying out experiments with the "little engine" at Soho, which it will be recalled was 18 in. diameter of cylinder by 5 ft. stroke, and would have liked to have "made haste slowly", but he was hustled by Boulton into making a big jump by designing two large engines, one of 50 in. diameter of cylinder for Bloomfield Colliery, near Tipton, Staffordshire, and the other 38 in. diameter of cylinder for John Wilkinson (1728–1808), the well-known ironmaster, for blowing his blast furnace at New Willey, near Broseley, Shropshire. Incidentally, we must mention that Wilkinson had, as Watt stated in a letter to Smeaton, "improved the art of boring cylinders so that I promise upon a seventy two inch cylinder being not farther distant from absolute truth than the thickness of a thin sixpence [say 0·05 in.] at the worst part". This boring mill was patented by Wilkinson in 1774; it consisted essentially of a rigid bar along which the boring head traversed through the cylinder, which was fixed. Previously the boring head had simply been forced through the cylinder centring itself by the interior. The advent of this boring mill was most opportune; we might almost say that it was vital to the success of the new engine, for not only did it bore more accurately diametrically, but most important of all, it bored the whole length of the cylinder truly cylindrical, a feat not achieved with earlier boring mills.

The Bloomfield engine was started on May 8, 1776, and the New Willey engine about the same time. Fortunately for Watt's reputation, and as if to justify Boulton's hustle, both engines were successful. The superiority of Watt's over the atmospheric engine, or as we shall now call it the common engine, in power and in economy of fuel—the consumption was about one-fourth —was demonstrated to the satisfaction of all enquirers. Enquiries flowed in from all quarters, particularly, towards the end of the year, from Cornwall where economy of fuel in pumping established a definite limit to the depth at which mining could be carried on and made to pay. Very soon that county became the principal sphere for exploitation of the new engine.

Watt and Boulton decided that their engine should be paid for by a royalty, or as they termed it a "premium", based very appropriately on the saving in fuel effected by the engine as compared with the consumption of a common engine doing the same work. Boulton and Watt stipulated that they should receive one-third of the value of the fuel saved; this was less than the actual saving, but they wished to be on the safe side. This premium was obviously greatest where coal was dearest, hence the importance of Cornwall as a sphere of operations. Since the simple case of a replacement of a common engine by the new one to do the same work hardly every occurred, Watt made elaborate tests on existing engines, both in Cornwall and in other places, to establish a relationship between his engine and the common one. Indeed, to him must be given the credit for the introduction and systematic application of tests of engine performance. Based on such trials as these Watt concluded that the effective piston load for the common engine was 7 lb. per sq. in., and for his own 10·5 lb.

To ascertain the saving, Watt compiled tables based on his figures giving a coefficient or constant for an engine of any diameter and stroke. As regards the fuel, it was the usual practice to keep an account of the consumption of coal at the mine, in order to claim remission of the duty, and the only unknown factor, therefore, was the number of strokes made. To register this, Watt introduced the engine counter, a form of reversed escapement actuating a train of counting wheels by the see-sawing of the engine beam upon which the counter was fixed in a locked-up box. All this was extremely scientific and entirely after Watt's own heart, but the counter could not differentiate between the different lengths of the strokes made, and this led to some justifiable grumbling.

Watt's single-acting pumping plant in its complete form must now be described, and this can best be done with the assistance of Fig. 13 which represents the engine for mine drainage as supplied in 1788, i.e. when the design had been more or less standardized. The boiler is of the waggon type, to be described later. Steam from it was led by a pipe to the top of the cylinder, and was there controlled by the steam valve. The latter, together with the other valves, was of the drop type, which Watt found easier to keep tight than either the sector valve or the plug cock of Newcomen.

It will be remembered that Watt's great object was to keep the cylinder as hot as the steam, and to effect this he enclosed it in a casing, which was in communication with the boiler. From the cylinder, a pipe, in which was the equilibrium valve, led from the top to the underside of the piston. In another pipe leading from the bottom of the cylinder was situated the exhaust valve, leading to the condenser. The latter was a cylindrical

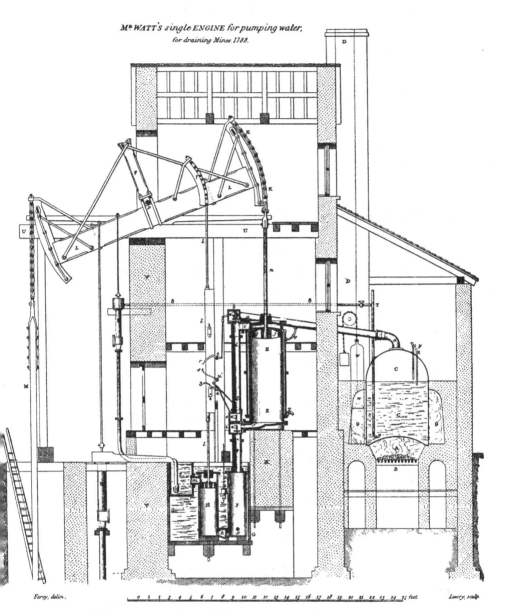

Fig. 13.   Watt's single-acting pumping engine, 1788–1800.

From Farey's *Steam Engine*, 1827.

vessel provided with an internal jet of water for condensation. Joined to the condenser was the air pump, actuated by the beam above, to withdraw the condensate and the small amount of air that came out of the water or leaked in through the joints. Both condenser and air pump stood in a wooden tank known as the hot-well. As the top of the cylinder was now closed, a stuffing box had to be devised for the piston rod to work through; its crosshead was connected by pitch chain and an arch head to the engine beam, which was stiffened by trussing in an elegant manner. The other end of the beam was directly connected by a similar chain to the pump rods. A plug-rod attached to the beam on a similar arch head had catches to engage with the three valves mentioned above.

Let us suppose that the engine is in operation and that the piston has arrived, as shown, at the top of its stroke. All the valves are open and steam is blown through all parts to warm them up and expel atmospheric air. The equilibrium valve is now closed, and the jet on the condenser turned on, forming a vacuum beneath the piston, while boiler steam, of roughly atmospheric pressure on the upper side of the piston, forces it to the bottom end of its stroke. The steam and exhaust valves are now closed and the equilibrium valve opened, thereby ad-mitting boiler steam to both sides of the piston. Seeing that it is now in equilibrium, the pump rods by their own weight bring the piston back to the top of its stroke. A pause now ensues the time of which is determined by a cataract, a form of governor that Watt found already in use in Cornwall; it consisted of a balanced tank into which water from an adjustable orifice flowed until the water capsized the tank and by a lever closed the equilibrium valve when the cycle recommenced. The first two strokes of the valves were worked by hand, requiring skilled manipulation, and the plug-rod then took charge of this duty.

As regards the novel features in the engine, the stuffing-box never seems to have presented any difficulty, but the packing of the piston engaged Watt's inventive mind for many years—indeed, the problem is still with us—but eventually he arrived at a satisfactory method which consisted in a recess in the edge of the piston, filling the recess with hemp soaked in tallow and squeezing the hemp down with a "junk" ring by means of studs.

A word or two about the construction and erection of the engine is necessary. Boulton did not, as he had foreshadowed in 1769, build a special manufactory for the engines, but adhered to the practice followed with the common engine, that is to build it on the spot, except that the firm made the specially intricate parts like the valves (or "nozzles" as they were called), supplied sets of drawings and sent an erector to the job to put together the materials provided by the client; in fact, the firm acted somewhat like a consulting engineer does to-day. Many of the materials were obtained locally, but special parts came from places as far apart as Cumberland and beyond London. John Wilkinson made nearly all the cylinders at Bersham, up to 1795. With the limited means of transport then available —usually by sea—this involved a great deal of organization, in spite of which delays were frequent.

Watt was in the flood tide of the installation of his engines in Cornwall when he got a shock which upset him very much, and that was the news that Jonathan Hornblower, son of the great Jonathan, whom we have mentioned as the "big noise" in Cornwall when Watt arrived there, had brought out an engine which was going to supplant Watt's engine entirely. Hornblower's idea, patented in 1781, was to use the steam successively in two cylinders, using part of its pressure in the first and part in the second cylinder, a method that is called nowadays "compounding" or "two-stage expansion". The engine consisted of two single-acting cylinders, the low pressure being in the usual place acting on the arch head at the end of the beam, and the high-pressure cylinder between it and the beam centre, acting on an intermediate arch head. Hornblower erected an engine at Radstock Colliery, near Bristol, for a Mr Winwood in 1782. William Murdock (1754–1839), the trusted outdoor assistant of Boulton and Watt, was sent to spy out the land and made a sketch of it, which he sent to Watt, and this is still preserved in the Boulton and Watt Collection. Watt was "much vexed" and called the Hornblower family hard names such as "Horners",

"Trumpeters" and "horned imps of Satan", but he was relieved when he found that the engine was not such a wonderful one as had been stated; an important fact was that it embodied, in a disguised form it is true, his separate condenser. The firm consequently informed the colliery proprietor that it was an infringement of their own engine and that if persisted in they would bring an action against Hornblower. However, the engine did not attain any particular success and Hornblower was left alone. He did, however, erect another engine at Tincroft Mine in Cornwall in 1790, and this was more successful. The truth of the matter is that with the low boiler pressure then available it was an unnecessary complication to have two cylinders, and consequently the two proved less economical than a single one. Hornblower applied to Parliament in 1795 for an extension of his patent; Boulton and Watt opposed the bill by vigorously lobbying and he failed to get an extension; the bill was thrown out and nothing more was heard of the engine.

The rotative engine. Almost from the start, the idea of applying the new engine to driving mills where rotative and not simple reciprocating motion as in pumping is required, was present to Watt's mind and indeed to that of many other persons. Almost as early as the time when he invented the condenser he proposed to get rotary motion direct by steam pressure in a "steam wheel" or rotary engine, but he never managed to make a success of it, although he actually made an experimental one. Incidentally it may be remarked that this method—so desirable theoretically—has been the obsession of hundreds of inventors since Watt's day. The other way of obtaining rotative motion, albeit indirect, was by harnessing the reciprocating engine, an idea present in Watt's mind as early as 1774 but reluctantly put on one side until 1781 because of his immersion in the task of building pumping engines. The *direction* of rotation has always been clockwise; this convention is possibly based on the way a man works a winch handle.

Boulton sensed the situation and was most insistent that Watt should produce a rotative engine, for whereas the field for pumping engines was comparatively restricted, Boulton realized that the field for rotative engines was almost illimitable. In 1781 we have Boulton writing: "the people in London, Manchester and Birmingham are *steam mill mad.*" Watt, who shortsightedly believed that pumping engines still offered the most remunerative field, complained as late as 1782: "surely the devil of rotations is afoot." The time was more than ripe. The need was so urgent that many engineers were trying to make the common engine work rotatively, and succeeded as explained in a previous chapter.

The problem that Watt had to face required inventive talent of a high order. Many changes in the mechanism were involved: steam should, for convenience and efficiency, be made to act on both sides of the piston; consequently, the piston rod must be coupled inflexibly to the beam, and at the other end of the beam a connecting rod or like mechanism must be adopted to turn the shaft; also a flywheel had to be added to carry the crank over the dead centres. Over head and ears in business as he was, and suffering from indifferent health as he did all his life, Watt attacked these problems, as well as that of expansion of steam, with brilliant success; indeed, this was the zenith of Watt's inventive career. Nothing could be simpler, one would think nowadays, than to apply the crank to the steam engine, but with the knowledge that the stroke was by no means invariable, it was believed that this would result in a smash, not realizing that the crank would control the stroke of the connecting rod. The crank was in common use, for example, in the spinningwheel and the foot lathe, and it is pretty certain that Watt did not consider its application to his engine to be patentable. However, in 1779, Matthew Wasborough of Bristol had made for James Pickard of Birmingham a common engine to work rotatively by a "rick-rack", i.e. a ratchet-and-pawl mechanism. It did not work well, needless to say, and Pickard replaced it by a crank and connecting rod. In 1780 he patented an arrangement for carrying the crank over the dead centres, consisting of a pinion gearing into the crank disc and having a weight on its circumference. It is this idea that all the evidence goes to show was stolen from Watt by his workman, Richard Cartwright.

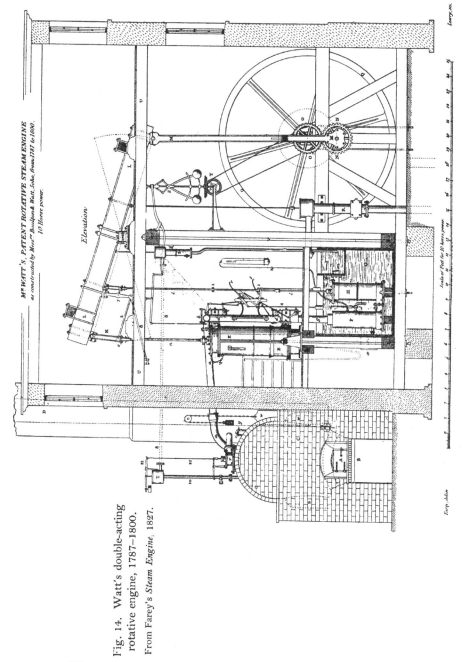

Fig. 14. Watt's double-acting rotative engine, 1787–1800.

From Farey's *Steam Engine*, 1827.

M.ʳ WATT'S, PATENT ROTATIVE STEAM ENGINE

as constructed by Mess.ʳˢ Boulton & Watt, Soho, from 1787 to 1800.

10 Horse power.

Elevation

The merit of Wasborough and Pickard was, in the author's opinion, that they had the enterprise and boldness to apply the crank, and the acumen enough, as engineers, to include the use of a flywheel.

However that may be, Pickard's patent seems to have been assumed to cover the application of both crank and flywheel. Rather than attempt what would have been a fairly easy task, viz. to upset Pickard's patent, the long-headed Watt realized that it would be bad policy on the part of one like himself, who held so many patents, to try to do this, lest someone should give him tit-for-tat. Watt therefore devised substitutes for the crank,

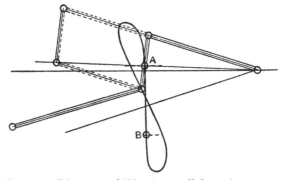

Fig. 15. Diagram of Watt's parallel motion, 1784.
Redrawn from the Boulton and Watt Collection, Birmingham.

which he patented in 1781. Only one of these—the "sun and planet" or epicyclic gear—came into use; in this a planet wheel, rigidly fixed to the end of the connecting rod, is made to move around the perimeter of a sun wheel keyed to the driven shaft; it is a property of the gear that if the wheels are of equal size, the driving shaft will revolve twice for every double stroke of the engine. This gear was used until Pickard's patent expired in 1794 and even later. It is to be seen on the engine depicted in Fig. 14.

The problem of an inflexible connection of the piston rod to the beam was solved in 1784 by the parallel motion included in Watt's patent of that year. It is a clever and elegant combination of the three-bar motion and of the pantograph (see Fig. 15). Long afterwards Watt said of it: "I am more proud of the parallel motion than of any other mechanical invention I have ever made."

To ensure steady motion of the engine, even when the load upon it varied, Watt, in 1787, introduced the conical pendulum centrifugal governor consisting of two balls which fly outwards when the speed increases and move a sleeve which by linkage controls a butterfly valve in the steam pipe. Watt very wisely neither claimed this as an invention of his own nor patented it, because it seems to be certain that something of the kind had already been applied to regulate the speed of mill-stones in flour mills. Watt knew of this but, as was always the case with him, he never touched anything but what he improved on it, and to this the governor was no exception.

The first double-acting rotative engine was put up in 1783 and it need hardly be said that it was for John Wilkinson, who was always keen on the latest novelty. The most elaborate engine was perhaps that of the Albion Mill at Blackfriars, London, 1784, which revolutionized the flour-milling industry, but was unfortunately burnt down in 1791. A rotative engine for winding ore from the mine, known as a "whim" engine, was put up in 1784, in Cornwall, by Boulton and Watt, and apparently some sort of gear-wheel train, analogous to what we use in the motor car, was used to reverse the motion. By 1795 we find Watt remarking that they "changed the motion by stopping the engine & setting the fly [wheel] agoing the contrary way". This was quite easy as long as steam was given to the piston during the whole length of the stroke. By about 1787 the rotative engine was standardized and is represented in Fig. 14, which shows it as it was made from that date until the expiration of the patent in 1800. An actual engine of date 1788 is preserved in the Science Museum, South Kensington. A new era was thus ushered in, enlarging enormously the field of application of the steam engine; indeed, it now entered upon a career of world-wide utility.

We have yet to mention an important discovery of Watt, viz. the principle of expansive working of steam. This he effected by cutting off the steam when the piston had completed only a portion of its stroke, and allowing the remainder of the stroke

to be made by the expansion of the steam, with resultant
economy. Watt had thought about this as early as 1769, and
had made actual experiments at Soho in 1776, but with the low

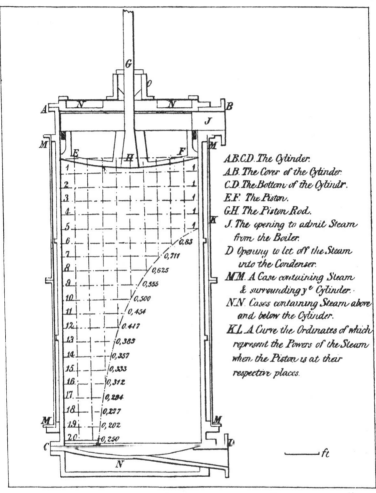

A.B.C.D. The Cylinder.
A.B. The Cover of the Cylinder.
C.D. The Bottom of the Cylindr.
E.F. The Piston.
G.H. The Piston Rod.
J. The opening to admit Steam
    from the Boiler.
D Opening to let off the Steam
    into the Condenser.
M.M. A Case containing Steam
    & surrounding ye Cylinder.
N.N. Cases containing Steam above
    and below the Cylinder.
K.L. A Curve the Ordinates of which
    represent the Powers of the Steam
    when the Piston is at their
    respective places.

Fig. 16. Watt's diagram showing pressures of steam expanding
in a cylinder.

From his patent specification, 1782.

pressure then used the advantages to be gained were small.
However, Watt included expansive working in his patent of
1782 with a diagram (see Fig. 16) which brings out clearly the
advantages of its use. It will be realized that this diagram is the

result of pure calculation. Watt lost interest in the idea because he saw clearly how small was the saving to be effected with steam at atmospheric pressure only, and the matter was not pursued further. After Watt's time, however, when pressures were increased well above that of the atmosphere, expansive working became all important for economy, as we shall see in a succeeding chapter.

As important to the steam engineer as the stethoscope is to the physician is the indicator, because it tells the tale of what is going on inside the cylinder. The original form of this apparatus, brought out by Watt probably about 1790 was, as its name implies, merely a pointer moving over a scale. The pointer was actuated by the pressure in a small cylinder which was in communication with the engine cylinder, and the pointer consequently reproduced the varying pressure there. The brilliant improvement, which transformed the indicator into a measurer of the work done in the cylinder, is believed to be due to John Southern in 1796. This improvement consisted in substituting for the pointer a pencil attached to the piston rod of the small cylinder, and placing behind the latter a tablet, carrying a sheet of paper, reciprocated by the engine itself; thus the pencil traced on the paper a closed diagram which showed what happened at every point of the stroke; by measuring the area of the diagram to scale the work done was known and the horse-power could be deduced from it. The instrument continued to be known by the old name, although "work measurer" or "ergmeter" would more correctly describe it. It was kept secret for years and entrusted only to the staff of the firm; indeed, it did not become generally known until John Farey found it in use by Boulton, Watt & Co.'s men in Russia in 1826, and brought back the instrument with him to England.

Inventors and infringers. It is hardly necessary to state that the success of Watt's engine brought into the field both inventors of other forms of engines, and also persons who sought to infringe or evade the patent. It is pertinent to observe here that Watt refused persistently to allow anyone to make his engine or any part of it under licence. This policy, coupled with the long period that the patent lasted, undoubtedly held back the development of the steam engine. Conspicuous amongst the inventors just named was Jonathan Hornblower, the younger, who had given Watt such a shock previously (see p. 78). Among the schemers was Edward Bull, who brought out in 1792 the inverted type of engine which goes by his name, assuredly in ignorance of the fact that this was the arrangement that Watt had schemed at the outset of his career. Bull's intention was to circumvent the patent, but in his engine the separate condenser was embodied, and it was therefore an infringement of Watt's patent. It was only after endless injunctions and litigation that in 1797 Bull was defeated in the Courts. It was not till 1799 that the validity of the patent of 1769 was incontestably established. This litigation, as is so often the case, was ruinous, and it was estimated to have cost the firm upwards of £10,000; another aspect of it is that it may be said to have diverted Watt's attention from what was his true vocation, i.e. invention. While the litigation was going on the Cornishmen were in open defiance, and refused to pay their premiums; this stung Watt to say of them to Boulton in 1796: "The rascals seem to have been going on as if the patents were their own.... We have tried every lenient means with them in vain, and since the fear of God has no effect upon them, we must try what the fear of the devil can do." This, in short, is what had to be done. With great difficulty, and not until some time after 1800, were most of the arrears of premiums, amounting to about £40,000, collected.

The partnership of Boulton and Watt, brought to a close in 1800, had been a happy one. Boulton's command of capital, his energy, tact and influence were complementary to the inventive genius and application of Watt. The enterprising, broad-minded, optimistic outlook of Boulton corrected the self-depreciatory, cautious and narrow outlook of Watt. After the latter's retirement from the cares of business, he mellowed wonderfully. Sir Walter Scott gives a delightful picture of him in 1817: "The alert, kind, benevolent old man, his talent and fancy overflowing on every subject, with his attention alive to everyone's question, his information at everyone's command." This is how he appears on the portrait that we give as a frontispiece. It was painted by John Partridge about 1806, when Watt was seventy years of age.

A new partnership was now formed, Matthew Robinson Boulton and James Watt junior, respectively taking their fathers' places. These two were young men with talent and energy, and with the assistance of servants of the old firm, William Murdock of the outdoor staff, and John Southern (1758–1815) the trusted assistant in the drawing office, entered upon a period of un-exampled prosperity. Soho became the nursery of the steam engine and of steam engineers.

The advances in constructional materials for engines made during Watt's time had been considerable. The single oak log of the atmospheric engine had been displaced by the compound beam of Smeaton and later by the trussed beam of Watt, only to be supplanted in turn about 1797 by the cast-iron beam. Timber had entered also into the framework, gearing and shafting; it, too, gave way to cast iron, so that by 1800 cast and wrought iron reigned supreme.

To sum up. The engine remained as Watt had found it—a vacuum engine, but with such improvements as constituted really a new engine. First as regards the pumping engine; he found the common engine as improved by Smeaton performing a duty of 7 to 10 million lb. of water raised 1 ft. high by a bushel of coal. At one bound Watt's pumping engine raised the duty to as much as 18 millions, and when fully established one engine did 26 millions. In 1792 Watt was able to report two engines in Cornwall doing the unprecedented duty of 35 and 32 millions respectively. By 1800 the firm was able to guarantee this as their everyday performance. As regards capital cost, owing to the way in which the pumping engine was put up by the client himself, we have no available data.

As regards the rotative engine, we have Boulton's statement that their engine would do as much work per hour with a bushel of coal (94 lb.) as would ten horses. With this engine the practice grew up of supplying it complete; the average customer did not want to be saddled with the trouble of getting materials together, nor had he any facilities for erection. We find that the capital cost of a 4 h.p. engine in 1788—the smallest made— was £354. 10s. As the premium hitherto charged based on savings of fuel consumed was inapplicable to a rotative engine, a fixed premium based on the horse-power was charged, viz. 6 guineas per annum in London or 5 guineas in the provinces.

The total number of engines built during the Boulton and Watt partnership was roughly 500, of which 38 per cent were pumping and 62 per cent rotative (mostly for the textile industry), showing that Boulton was eminently right when he advocated concentration upon the latter type. Averaging the engines at 15 h.p. each—a not improbable figure—we get a total of 7500 h.p. built in twenty-five years, a mere fleabite, when we reflect that to-day a single turbo-generator unit may be twenty times that size.

We are of the opinion that Watt's master patent for the separate condenser, with the extension granted to him by Parliament, amounting in all to thirty-one years, was unduly long in the public interest. It had tied down progress to the wheels of Watt's chariot, for it must be remembered that he frowned upon any increase of steam pressure beyond a few pounds above the atmosphere on the grounds of safety. With him enterprise had stopped; on the other hand, we have to remember that but for the protection afforded by the patent, Watt would not have enjoyed the environment in which he was able to work out his equally brilliant inventions, necessary to bring the rotative engine into being, and for this we must always be thankful. It had enabled him also with Boulton's help to secure some financial reward for his services, and eventually wide recognition from the public generally; in this respect he escaped the sad fate of so many great inventors of this great period in industry. Looked at broadly, it can be said that no one has made a greater contribution in ultimate effect to the development of the steam engine than has James Watt. Whatever the fate of the reciprocating engine, and it will be foreshadowed in succeeding pages, his separate condenser goes on fulfilling in the steam turbine as important a function as ever it did before.

Before concluding this chapter as bearing on the spread of the steam engine, a few remarks are desirable as to the phenomenal progress that had taken place in the century that was thus brought to a close. The canalization of rivers known as navigations, the digging of trunk canals and of docks, the improvement of turnpike roads, had led to more rapid transport and the economical carriage of bulky articles. Coastwise shipping, and its corollary local wooden shipbuilding, closed the links in this chain of transport. Coal and ore mining were being prosecuted to ever greater depths and drowned areas were being recovered. The smelting of pig iron had advanced by leaps and bounds owing to the introduction of coke as fuel and to higher blast pressures. This advance was correlated with the demand for iron castings and for pig iron for Cort's puddling processes, which had greatly improved the quality of wrought iron and had increased its output. The textile industry had been transformed from handicraft into a machine industry. Less spectacular but noteworthy were the changes in other industries such as hardware and pottery. The Enclosure Acts, while inflicting hardship on the peasantry, had encouraged land improvement and stimulated land reclamation; these in turn led to the beginnings of mechanization of the farm. Development is cumulative in effect and ancillary industries such as chemical manufacture sprang up. The effect of these upon engine building was to stimulate demand; latterly output had scarcely kept pace with requirements, owing to the slowing-down influence of Watt's patent.

Cornish pumping engine, 1811.

CHAPTER VI

LOW-PRESSURE AND HIGH-PRESSURE ENGINES, 1801–50

State of the art of engine construction—High-pressure engine—Trevithick's "puffer"—Evans's engines—Trevithick's locomotives—Alban's high-pressure engine—Woolf's compound engine and boiler—Trevithick's plunger-pole engine—Cornish engine—Taylor's engine—"McNaughting" engines—Side-lever engine—Grasshopper engine—Table engine—Horizontal engine—Oscillating engine—Machine tools—Details of engine—Valve gear.

T HE nineteenth century opened, as far as concerned the steam engine, with a clean slate. Not only had Watt's patent for the separate condenser expired, but his subsequent patents in which that invention was embodied had also run out. Everybody was free to build both pumping and rotative engines. The atmospheric engine was obsolescent, but by no means extinct, for it continued to be built by reason of its cheapness in first cost for a few duties like winding from shallow mines.

At the present day we are prone to wonder why these and other indirect types of engines, so clumsy-looking to our eyes, should have survived so long. Apart from tradition there was a sound, almost instinctive, reason for the survival, well understood by the old millwrights, and that was because the mechanical construction of those days necessitated it. Working parts might not be quite in line; bearings might not run quite true; cranks and other parts might not be quite square. What the millwright wanted was provision of means for adjustment of alignment and of centres by packing, by keys and cotters; he favoured long beams and rods of timber or of iron, because they admitted of some amount of elasticity or "give"; he kept speeds low for fear of bearings heating; he liked transmission by cogged wheels. Anyone who has watched an old engine working will realize that a certain amount of "slogger" necessarily takes place. This state of things arises from the fact that machine details had to be made by hand individually, and had to be designed accordingly. Castings were used as far as possible rough from the foundry; removal of excess material was reduced to a minimum because it had to be done by chipping and filing, consequently the smith forged to finished sizes; plane surfaces were avoided wherever possible by the use of linkage rods; cotters, not screws, were used for making adjustments; concentricity and squareness of wheels on shafts were ensured by "staking on"; little else but gudgeons were turned, as turning had to be done by hand tools without back gear.

Although the dead hand of Watt was removed, his tradition, amounting among engineers almost to a fetish, held sway and the rotative beam engine continued to flourish like the green bay tree and indeed in the first half of the nineteenth century was in its heyday. So much so that it became almost standardized, if we are to judge by the treatises of some writers on the steam engine, like John Bourne,[1] who published tables of dimensions of every detail for a range of sizes, as if engine building could be reduced to but little more skilled a trade than bricklaying.

However, the needs of industry were affording scope for new applications of engine power, fresh brains were at work, a flux of ideas set in, inventive talent seemed to be released. The developments that took place in consequence may be said to have taken place roughly in three directions and to have progressed contemporaneously: (1) the appearance of a number of new types of engines, (2) the rise of the Cornish engine and the compound engine from the Watt pumping engine, and (3) the invention of the high-pressure engine. The last-named development being a new departure altogether, we shall take it first.

The term "high pressure" connotes an engine which dispenses with the vacuum and works solely by the pressure of steam above the atmosphere, into which the steam is discharged after it has done its office. The conception was not by any means new, for it had occurred to many persons; for instance, Jacob Leupold (1674–1727) of Leipzig, Saxony, in 1724–7 published a description with a diagrammatic sketch (see Fig. 17) of such an engine.

[1] Bourne, *Treatise on the Steam Engine*, 1846, pp. 140–8.

Watt had the idea too, but dropped it. His assistant, William Murdock, in 1785 made a model of a high-pressure engine—quite a notable one—but it was suppressed at birth, for Watt would have none of it, and to the end of his days he was satisfied with the vacuum engine.

It is difficult at the present day to realize what fundamental changes were involved in introducing the high-pressure engine. The ponderous beam see-sawing slowly in a massive engine

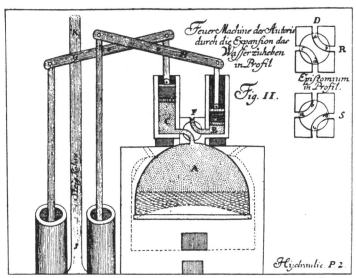

Fig. 17. Leupold's design for a high-pressure steam engine.

From his *Theatrum Machinarum Hydraulicarum*, 1725.

house crowded with valve gear, air pump and condenser, costly to install and run, was replaced by a faster-running direct-acting engine with a simple valve gear, occupying little space, requiring hardly any foundation, cheap in first cost and easy to work. Perhaps the comparison can be visualized most readily to-day if we think of the steam engine and its boiler as compared with the internal combustion engine of the same power. Moreover, the high-pressure engine lent itself to being made in small sizes and was fitted to perform the humblest task, even that of a few labourers; this vastly extended the range of industry that steam power could serve. It is true that sacrificing the vacuum meant that the engine was not economical in fuel, but its great advantages outweighed this consideration, particularly where fuel was cheap.

The pioneers in the development of this engine were Richard Trevithick (1771–1833) in England, and Oliver Evans (1755–1819) in the United States. Trevithick was brought up in the school of practical experience, and at the age of nineteen he was occupying responsible positions on the Cornish mines.[1] As suggesting how the idea of the engine arose in his mind, it may be mentioned that in 1797 he had introduced the water-pressure engine, several of which he erected. If we imagine high-pressure steam to be used instead of high-pressure water, we have the idea underlying the new engine. Whether it was in this way Trevithick's mind worked or not, we find him about 1798 experimenting with a model at Redruth. He realized clearly that he was sacrificing the vacuum and he therefore enquired from his lifelong friend and mentor, Davies Giddy (1767–1839), later known as Gilbert, and subsequently President of the Royal Society: "What would be the loss of power in working an engine by the force of steam raised to the pressure of several atmospheres but instead of condensing to let the steam escape?" Gilbert replied that "the loss of power would be one atmosphere diminished by the saving of an air pump with its friction and in many cases with the saving of condensing water". When Trevithick received this answer, Gilbert says: "I never saw a man more delighted."

No progress could be made, of course, without a suitable boiler capable of supporting a pressure of, say, 50 lb. per sq. in. The lack of such a boiler had been the stumbling-block in Savery's path, but boiler construction had made some advance since his time and Trevithick was not one to be daunted. His first idea was to make his boiler globular or pear-shaped, of cast iron, with an internal firebox, as shown in his patent specification; he soon arrived, however, at the cylindrical form with the back end dished and an internal U-shaped flue inserted from the

[1] See Dickinson and Titley, *Richard Trevithick, the Engineer and the Man*, 1934, for further details of his career.

front (see Pl. III). The cylinder was sunk in the boiler, where it would keep hot. To distribute the steam to the cylinder Trevithick employed a simple mechanism. This consisted of a tappet on the crosshead which struck up and then down a spanner or lever on a 4-way plug cock like Leupold's; the plug had passages in it such that it admitted steam first to one and then to the other end of the cylinder. With such a cock, steam must be admitted during the whole of the stroke. In 1810, for a single-acting cylinder, Trevithick devised a plug cock so that it would work expansively. The whole engine was eminently simple and compact.

We now leave Cornwall for the United States, where in Philadelphia Oliver Evans was carrying on his experimental work.[1] To visualize what was the stage to which the art of engine building had attained there, we should mention that in 1803 not more than six engines could be mustered in the whole of the States; mechanical construction and skill were at least fifty years behind those of England. Evans's engine, constructed in 1804, was direct double-acting vertical with a cylinder 6 in. diameter and 8 in. stroke (see Pl. IV). As may be imagined, it was largely the product of the carpenter and the smith, e.g. the crosshead was of hard wood and the flywheel, $7\frac{1}{2}$ ft. diameter, was built up of laminated boards. Steam admission and exhaust were controlled by 3-way cocks actuated by adjustable straps from pins in the flywheel. As in the case of Trevithick's engine, Evans relied on friction between the plug and its body to retain the former when its port was brought into coincidence with that in the body. The speed was about 30 r.p.m. A disadvantage was that any variation in speed was liable by inertia to overthrow the plug, so that the ports did not match.

The boiler was noteworthy, for it consisted of a copper shell 15 ft. long by 36 in. diameter tapering to 30 in., closed at each end by cast-iron headers through which passed a fire tube strengthened by iron rings. The exterior of the shell was lagged with wood over which iron hoops were driven. The engine was

[1] See Bathe, G. and D., *Oliver Evans—A Chronicle of early American Engineering*, for further details of Evans's career.

PLATE III. TREVITHICK'S HIGH-PRESSURE ENGINE AND BOILER
BY HAZELDINE & CO., 1805

Preserved in the Science Museum, London

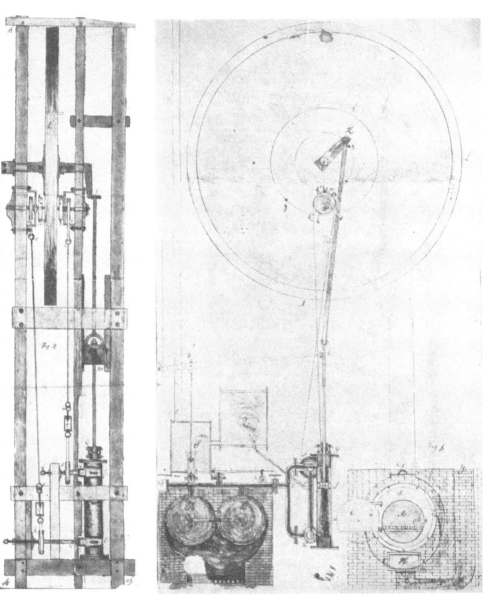

PLATE IV. OLIVER EVANS'S HIGH-PRESSURE ENGINE AND BOILER, 1804
From the original drawing in the Transportation Library, University of Michigan

a success, for Evans was able to write:[1] "I succeeded perfectly with my little engine. I could break and grind 300 bushels of plaster of Paris, or 12 tons in twenty-four hours; I applied it to saw stone in Market Street [i.e. Philadelphia], where the driving of twelve saws in heavy frames, sawing at the rate of 100 feet of marble in 12 hours, made a great show."

We have described this little plant at some length because it embodied ideas which were used, as we shall see later, in subsequent developments, developments in which Evans was not destined to take part for the reason that he received practically no encouragement; indeed, he was derided and criticized for his attempts. Like Trevithick, he was ahead of his time; he was a pioneer and deserving of a place with other honoured names on our frontispiece. This portrait, representing him at the age of fifty, is reproduced from an engraving by W. G. Jackman, artist unknown, published in 1862.

To Trevithick we must now return. Not content to exploit his high-pressure engine, or "puffer" as it was nicknamed, he realized that he had at hand for the first time a power that could be applied to move a vehicle on a common road. To the task of bringing into being a steam carriage he now addressed himself, and soon had one ready for the road. We have no account of the difficulties he had to overcome, formidable as they must have been. All we know is that on Christmas Eve, 1801, the carriage was tried on Beacon Hill, a slight up-gradient out of Camborne, his native town. Why he should not have chosen a level road is a mystery. The carriage failed for want of steam, but the day after Boxing Day, December 27, he ran it again. This time it suffered damage owing to the roughness of the road; in consequence, as Giddy informs us, it was dragged into a shed while the party consoled themselves with "a roast goose and hot drinks" in a neighbouring hostelry. "Forgetful of the engine, its water boiled away, the iron became red hot, and nothing that was combustible remained either of the engine or the house." What a jovial crew they were!

Trevithick and his friends were sufficiently satisfied with this

[1] *Niles's Weekly Register* for 1813.

success for him to be sent to London to obtain a patent, which he did. In this epoch-making patent, March 24, 1802 (No. 2599), he covered both his high-pressure engine and the steam carriage, as the title indicates: "Steam engines—Improvements in the construction thereof and Application thereof for driving Carriages."

In 1803 he built another steam carriage and ran it in London to advertise it. We know very little about it except that it ran several trips from Leather Lane, Holborn (where it was put together), to Paddington, a distance of some four miles, and back. No one seems to have taken any notice of the event; the fact was that neither the public nor the road surfaces were ready for the innovation. Trevithick reluctantly but wisely decided to drop the experiment and devote his attention to the exploitation of his stationary engine, which was meeting with a great deal of success. In pursuance of this aim he was down in South Wales erecting his engines for rolling mills at different ironworks. The proprietor of one of these works, Samuel Homfray (1761–1822) of Penydarren, a great sporting man, bet Anthony Hill, his neighbour at Plymouth Ironworks, 500 guineas that he would produce a locomotive to run on the cast-iron tramway then in use connecting their respective iron works with the Glamorganshire Canal. It is hardly to be doubted that Trevithick had surreptitiously put Homfray up to make this bet. It was accepted. Trevithick got to work at once—nothing pleased him better than to start a fresh hare—and in a few weeks' time he had knocked together a locomotive that made a successful trip on the tramway from the works to Navigation House on the Canal, a distance of $9\frac{3}{4}$ miles, not, however, without causing several breakages of the tram plates. One thing in his favour was that the gradient was a descending one! The date of this trip was February 21, 1804—a day ever to be remembered in the history of the rail locomotive. The bet was won and that for the moment was all that mattered. No one realized the underlying import of the new method of transport; no one gave Trevithick any encouragement to persevere in overcoming the defects that had been revealed by the experiment. There the matter rested.

He fell back upon the stationary engine, and in order to put himself in funds to carry on the exploitation of his patent he was obliged to sell some share in it; one of the incoming partners was Homfray himself, who was alive to the value of the engine for stationary use in ironworks like his own. The policy of granting licences under royalty to firms up and down the country to make engines under Trevithick's supervision was continued. From a letter of Trevithick, dated September 23, 1804, to Giddy, reporting progress, we learn that close upon fifty engines had been supplied for such varied purposes as driving sugar mills, grinding corn, pumping water, winding coal and rolling iron. The engines that he was building were like those of which we have already given an illustration (see Pl. III), but some were horizontal. The field in which steam power could be made applicable was extended widely and the demand was commensurate.

An unfortunate accident with one of his engines at Greenwich on September 8, 1803, when the boiler blew up, caused much misgiving on the part of prospective customers, misgiving that was exploited by Boulton, Watt & Co., and other condensing engine-makers, who made capital out of it with that hoary phrase "I told you so", despite the fact that the accident was traced to the recklessness of a labourer who fastened down the safety valve while he was having a meal. Trevithick, on the other hand, reacted typically to the accident by deciding to make the boiler fool-proof. He wrote: "I shall put two steam [i.e. safety] valves and a steam gauge in future so that the quicksilver shall blow out in case the two Valves sh'd stick." Not only did he do so but he originated the lead rivet in the top of the boiler furnace, so that if the water there got so low as to uncover it, the rivet would melt and let out the steam without doing damage —the precursor, in other words, of the safety fusible plug.

We should like to follow up Trevithick in his next exploit— that of running a passenger locomotive in London—but we must desist and simply say that there has been no discontinuity between his various experiments and the development of the locomotive as we know it to-day.

There were many other engineers who realized that only in high pressures of steam was the source of greater economy in fuel to be found, and one of these who deserves particular mention is Jacob Perkins (1766–1849), an American inventor who came to England in 1819 to introduce his new method of engraving to stop the forgery of bank notes. He experimented with steam up to as high a pressure as 1500 lb. per sq. in. and examined its properties. In 1827 he patented (No. 5477) and constructed a single-acting compound engine in which the flow of the steam was uni-directional, i.e. it was admitted at one end and exhausted at the centre of the cylinder. He used a second cylinder like the first.

Another of the inventors who took up high pressure was Dr Ernst Alban (1791–1846), of the Grand Duchy of Mecklenburg-Strelitz. His bent was mechanical, but so little was known of engine construction there at that time that his parents destined him for the Church, and after that for medicine, before he was allowed to give himself to his bent, that of mechanics. His engine of 1828 had two single-acting opposed plungers, something on the lines of Trevithick's plunger-pole engine but horizontal. His 10 h.p. engine had plungers 3 in. diameter by 18 in. stroke, working at 650 lb. pressure. One of his early engines is preserved in the Deutsches Museum, München.

The difficulty with all these proposals for using high-pressure steam was that of getting boilers to stand them with safety. Naturally there was the usual prejudice against any new idea, and with justification, for not only was boiler-making merely an adjunct of the trade of the smith, but it was not thought necessary for the stoker placed in charge of a boiler to have any more skill than that of a common labourer. However, the fillip given by the high-pressure engine had its effects on the boiler, but a discussion of this must be reserved for another chapter.

We have now to hark back to the beginning of the century to trace the progress of the Watt pumping engine after Boulton and Watt had left Cornwall to its devices; this led eventually to the rise of the Cornish engine, and as the compound engine was also involved in this development we must consider them together.

In the previous chapter it was shown how the idea arose of making an engine in which two cylinders successively should be used and why it did not answer because the then range of steam pressure was insufficient to justify it. However, the idea only lay dormant, and now that it was practicable to attain higher pressures the time was ripe for further progress. In 1803 Arthur Woolf (1776–1837) patented (No. 2726) and brought out such an engine, together with a boiler (see p. 127) to supply it. Woolf had been in a subordinate position on a Cornish mine. He had come up to London, where eventually he had obtained employment as engineer to Meux's Brewery, in Tottenham Court Road. He received permission from the proprietors to experiment with an engine there by adding a high-pressure cylinder to it. He had made some experiments on the expansion of steam and whether his data were wrong or his deductions were unsound, the fact is that he got it into his head that, as set forth in his patent specification of 1804 (No. 2772), steam of 20, 30 and so on pounds pressure per sq. in. will expand 20, 30 and so on times its volume and still be equal in pressure to that of the atmosphere.[1] This was unfortunate, for he made his high-pressure cylinder too small, i.e. one-eighteenth of the volume of the low-pressure cylinder, and the engine would not do the work required. He tried all manner of ingenious expedients except the right one; after two years of experiment, his employers got tired of him and called in John Rennie—not altogether a disinterested person for he was a Boulton, Watt & Co.'s man—and he found that the performance of Woolf's engine as compared with one of Watt's was in the ratio of 3 : 4; Rennie was of the opinion that the latter "was better and simpler and easier of construction than Woolf's". In disgust, Woolf threw up his job at the brewery and entered into partnership with Humphrey Edwards, a millwright, who had a workshop in Mill Street, Lambeth. After some years they arrived at a standard design. The cylinders, instead of being arranged as had been done by Hornblower, were side by side with the two

---

[1] Cf. *Trans. Newcomen Soc.* XIII, p. 55. He had not realized even in 1827 that his ideas were erroneous, see p. 63.

pistons attached to the same pin of the parallel motion; the ratio of the cylinder diameters was 1 : 5. A comparative trial in 1811 with a Watt engine is said to have shown an economy over it of about 50 per cent in fuel.

Woolf, who was a difficult man to work with, dissolved partnership with Edwards in 1811. The latter, after a few more years in Lambeth, having established a market for his engines in France, and seeing a future there, migrated to Paris. John Hall of Dartford continued to look after the manufacture of the engine and boiler in England and made a number of them for export. Edwards obtained a French patent for the engine on May 25, 1815; eventually he joined Périer at the old Chaillot Foundry under the style of Scipion Périer, Edwards & Chappert. Edwards took the Woolf engine with him to Chaillot and went on making it with hardly any change for the rest of his life. It is stated that he put up on the Continent about a hundred engines made in Lambeth and, after his migration, about 200 made at Chaillot. Other firms took up the manufacture and such engines were being made up to about 1870. Thus it is explained why on the Continent the compound engine is usually called a Woolf engine. The main reason why the engines were so extensively made there and not in England was the relatively higher cost of fuel and hence the greater incentive to economy abroad than in England.

The reason why Woolf dissolved partnership with Edwards was probably because he had heard of the revival of interest in the performance of the engines in the mines in Cornwall, and thought there would be an opening there for his compound engine. Whatever the reason was we find Woolf back in Cornwall in 1811, in charge of engines at Crenver, Oatfield and Wheal Abraham. Now this revival was due to the lamentable fact that the performances of the engines had retrograded since Boulton and Watt had left Cornwall in 1800; whereas the duty of the Watt engine as they left it had been about 30 million lb. per bushel of coal, in December, 1811, the average with twelve engines was found to be only 17 millions. There were far-sighted engineers who deplored this falling from grace, among them Captains John Davey of Gwinear and Joel Lean of Crenver. These Captains concluded that the best way to improve performance was to publish reports of what was being done and thus excite emulation. Captain Lean was asked to take upon himself the office of Reporter and Registrar of the engines; he did so and on his death was followed by his sons, Thomas and John.[1] It is probably not too much to say that the practice of reporting was attended with more benefit to the mining industry than any other single event, the introduction of the Watt engine only excepted. As showing what a progressive improvement resulted as the steam pressure went up, it may be mentioned that the average of the largest, i.e. the most efficient engines, improved from a duty of 27 millions in 1814 to 55 millions in 1835; the improvement continued during the whole period of the Cornish engine, to be referred to shortly; in 1842 Taylor's well-known engine on the United Mines, Gwennap (see Fig. 18), reached its highest performance, giving a duty of 128 million ft. lb. per cwt. of coal,[2] but even higher figures were reported.

About the same time that Woolf returned to Cornwall, so likewise did Trevithick. After a hectic time in London, first on the Thames Driftway, then on dredgers, iron tanks for cargo stowage, floating docks and salvage operations, in 1810 he had a very serious illness, and in 1811 he was adjudged bankrupt. He lost everything and consequently retired to Cornwall to lick his sores and start life afresh. He had sold the last remaining share in his engine patent, but his mind still ran on high-pressure steam. He was a man of abounding energy who could not be idle long, and soon he was busy on pumping engines. We surmise that his hydraulic engine was in his mind and he conceived the idea that he could make the plunger of this engine work by steam. We imagine this because the engine he brought out, known as the "plunger-pole", briefly described, had a plunger instead of a piston. He built several of these engines

---

[1] Cf. Lean, Thomas, and Brother, *Historical Statement of the...duty performed by the Steam Engines in Cornwall*, 1839.

[2] The weight of the bushel as used in the Reports of the Lean brothers, *loc. cit.* p. 140, was 94 lb., so that the comparative figure is 107·4 million.

Fig. 18. Taylor's Cornish engine on the United Mines, 1840.

From the Perran Foundry Catalogue, 1868.

between 1812 and 1815, notably one at Herland Mine in the latter year, and got very high duty from it, more perhaps because of the high pressure he employed than because of the merits of the design, for the large surface of plunger exposed to the air on the out stroke must have wasted much heat. Not only did he build new engines but he added plunger poles to existing engines, accompanied of course by new boilers to attain the higher pressure demanded; in effect this amounted to compounding the engine. He did not stop there, however, for he pushed up the pressure of the single-acting engine, and at Wheal Prosper in the parish of Gwythian in 1812 he erected an engine which embodied the features of the Cornish engine, that is to say, a condensing engine of Watt type but using high pressure and working the steam expansively. The introduction of the Cornish engine can therefore be claimed for Trevithick. However, he was not destined to follow it up; his attention had been switched off to mining in Peru and to riches beyond the dreams of avarice that awaited there anyone who could drain mines at high altitudes, a task for which an engine using atmospheric pressure was useless. This was just the opening for his high-pressure engine, and Trevithick set sail for Peru in 1816. He left his interests in charge of Captain James Sims, who continued to build the plunger-pole engine for a few years longer and then quietly dropped it. Our portrait of Trevithick (see frontispiece) is reproduced from an oil painting taken at this turning-point of his career when he was forty-five years of age; it depicts a decidedly virile and arresting personality. The portrait, by John Linnell, is in the Science Museum, London.

Meanwhile Woolf had put up his compound engine, in 1814, at Wheal Abraham. The boiler pressure was 40 lb. per sq. in. and the expansion 8 to 9 times. In 1815 he put up a similar engine at Wheal Vor and both did extremely well. It was found, however, after prolonged working, that the original efficiency was not maintained, and soon the enthusiasm of the adventurers who had adopted his engine began to wane. In order to make a comparative test, in 1824 Woolf was instructed by John Taylor to set up at Wheal Alfred two engines, one of his own with his own boilers and the other a single-acting engine with Cornish boilers. The engines began working in the autumn. The latter engine was found to give a duty of 42 millions as against 40 millions from Woolf's engine.[1] What Lean says sums up the matter:[2] "The advantage which these engines possessed over those of a low pressure was soon made known by the monthly reports. The greater expense of their erection and the want of simplicity in their construction, were objections to their general use." The result was that Woolf decided to drop his engines and this was the last one that was made. Thus, curiously enough, the compound engine died out in England for the second time. Henceforward Woolf devoted his talents to perfecting the Cornish engine and became the most prominent engineer in Cornwall; in 1828, of the sixty engines entered in the *Monthly Reports*, seventeen were under his care. Nor was this all. As Lean again says:[2] "Besides being an experienced engineer, Woolf was a skilful workman; and the engines erected under his superintendence excelled in correctness of construction. After his example or by his instructions, other workmen also attained perfection in the art; and the engines made in Cornwall were found to yield in excellence to those of no manufactory however eminent." Woolf was a partner in Hayle Foundry, the oldest of these Cornish engineering works established in 1775 or thereabouts. Another celebrated works was Perran Foundry, established in 1791, and at its zenith between 1840 and 1860. The latter closed down in 1879, but Hayle Foundry lasted until 1903. There were other establishments, likewise gone alas, that owed much to Woolf's influence. The tradition was carried on by a band of men, William Sims, Michael Loam, John Taylor, William West, William Michell and others, whose productions went out into all countries where mining is carried on.

As to the cycle of the Cornish engine (see Fig. 18) we can only say that it was practically identical with that of the Watt engine; the high pressure and early cut-off were what caused such a superiority in the performance. The Cornish engine had

[1] *Newcomen Soc.* XIII, p. 63.
[2] *Reports, loc. cit.* p. 152.

one great advantage, and that was that when the sinking was going to be of considerable depth, the engine could be designed to be suitable for the heaviest duty it would be called upon to perform. At the beginning of the sinking, the cut-off would be early, and altered to be later and later as more and more pump rods were added to the load. Suffice it to say that under the charge of the engineers whom we have mentioned above, the Cornish engine created higher and higher records. Perhaps the best known is William West's engine at Fowey Consols, which in 1835 attained the unprecedented duty of 125 million ft. lb. per bushel of 94 lb. of coal. Although such remarkable duties were obtained with it, the performances were regarded outside Cornwall with scepticism. Nevertheless they could not be ignored, and Thomas Wicksteed (1806–71), who was favourably impressed by reports on the engine, in 1838 was instructed by the directors of the East London Waterworks Co. to go into the question. An engine 80 in. diameter by 10 ft. stroke, the counterpart of the one at Fowey Consols, was erected at the Waterworks at Old Ford, and tried against a Boulton, Watt & Co. engine 60 in. diameter and 7 ft. 11 in. stroke, when the former was found to give a duty $2\frac{1}{2}$ times that of the latter.[1] Pumping heads at waterworks are almost always less than those of mines, and the weight of pump rods is consequently less; it was found that waterworks engines ran at a higher speed than mine engines. The sudden closing of the valves, which were at first of the flap type ordinarily in use in Cornwall, gave rise to considerable shock; this defect was overcome by replacing the valves by the double-beat valve patented (No. 8103) by Nicholas Harvey and William West in 1839. This was really a modification of the steam valve of Hornblower of 1781.

As an outcome of Wicksteed's experiments, the use of the engine spread to other districts, for mines as well as for waterworks duties. Existing Watt engines were even converted to use high-pressure steam, new boilers being installed. The size of the engine increased, and in 1857 one of 112 in. diameter

[1] Wicksteed, *An experimental enquiry concerning . . . Cornish and Boulton and Watt pumping engines*, 1841.

cylinder and 10 ft. stroke was erected at the Southwark and Vauxhall Waterworks pumping station at Battersea. The steam pressure had, by 1850, been raised as high as it was possible to expand economically in a single cylinder. The limit was reached; hence the engine has been superseded by other types, but Cornish engines are still in service and it can be truthfully said that they never wear out. Space will not admit of following the pumping engine further.[1]

The compound engine, which had continued to be used on the Continent ever since 1815 although practically forgotten in England, now reappeared in England quite independently and one might say by the back door. It was in this wise. There were many mills and factories which, as time went on, experienced the need for more power. The existing beam engines—still the accepted type for such service—could not take more load. The problem of how to increase the power, short of scrapping the plant, was solved in 1845, by John McNaught of Bury, Lancashire. He compounded existing engines by the addition of a high-pressure cylinder at the other end of the beam, i.e. between the trunnions and the crankshaft (see Fig. 19). This was nearly always possible as the factor of safety in these old beam engines was very high and the additional stress involved did not endanger the engine. This practice, known as "McNaughting", had quite a vogue and had some advantage in relieving the stress on the beam centre, so much so that his plan was adopted in many instances for new engines. Compounding never thereafter dropped out of the scheme of things.

McNaught was a prolific engineer, and we may mention that between 1825 and 1830 he simplified and improved the engine indicator, lightening the parts to reduce distortion and making it suitable for higher speeds and greater pressures. He surrounded the small cylinder, in which was the spring-controlled piston, by a casing on which he placed a sheet of paper—or indicator "card"—the pencil hung down from the piston rod over the paper and the diagram was drawn thereon.

[1] But cf. Westcott, G. F., *Pumping Machinery*, Science Museum Handbook, 1932.

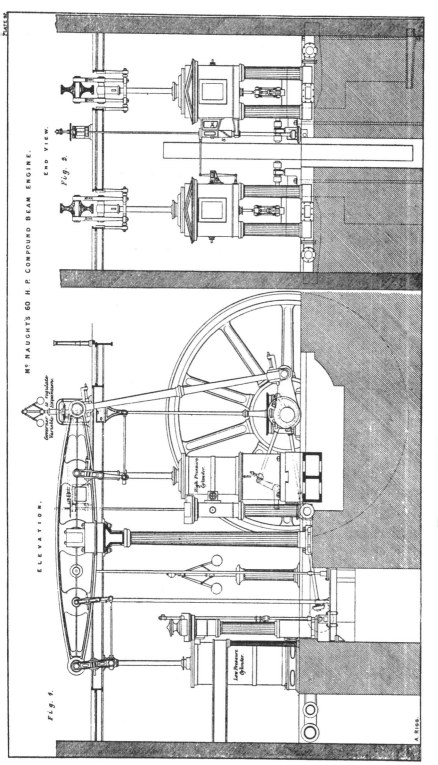

PLATE 57.

Mc NAUGHT'S 60 H. P. COMPOUND BEAM ENGINE.

END VIEW.

Fig. 2.

ELEVATION.

Governor to regulate Variable Expansion.

High Pressure Cylinder.

Low Pressure Cylinder.

Fig. 1.

A. RIGG.

Fig. 19. "McNaughted" compound beam engine.
From Rigg's *Steam Engine*, 1878.

We must now revert to the beginning of the century to consider the third line of development, that is, the introduction of different types of rotative engines. There were many of these, for inventive talent ran riot, but we can only notice those which experienced some length of service, or filled some want. We commence with the work of Matthew Murray (1765–1826), for he was the first and most considerable competitor of Boulton, Watt & Co. Murray came from Newcastle-upon-Tyne, and set up in Leeds in 1789. In 1805 he brought out a self-contained or "portable" engine, as it was termed, something like the old beam engine turned upside down. The cylinder remained vertical and the crosshead, by dependent side rods, worked the beam centred below; the other end of the beam, by an upwardly-directed connecting rod, worked the crankshaft above. This engine is mentioned, not to show, although it does so, how difficult it was for engineers to shake off the idea that a beam and upright cylinder were necessarily parts of an engine, but because this type was taken up by other makers under the name of the side-lever engine, and remained for about forty years the established type for marine-engine practice; the great advantage of the arrangement was that the centre of gravity was brought well down in the ship. Very few, however, were made for land service.

Another type that had considerable vogue was the half-beam or grasshopper engine. We find this engine patented in 1803 by William Freemantle, but the credit for its actual introduction must be assigned to Oliver Evans, who independently brought out his "drop valve" engine embodying this linkage in 1804. The cylinder is still vertical and the piston rod is connected to a beam roughly four times the length of the stroke. The centre of this beam is constrained by a bridle on either side attached to the framing and thereby a straight-line motion is impressed on the crosshead. The other end of the beam is pivoted on two long back links which rock to and fro to suit, and it is this motion (see Fig. 20) that gives the engine the name grasshopper. It will be observed that the pressure on the piston is transmitted to the crank without passing through the beam gudgeon, so that some frictional loss is obviated. It may surprise engineers to know that a grasshopper engine was in use as late as 1870.

An engine in which the beam was dispensed with but indirect action was retained was the table engine, patented in 1807 by Henry Maudslay (1771–1831), so well known for his introduction of the slide rest into engineering workshops. The cylinder was still vertical, and carried on a "table" or platform supported

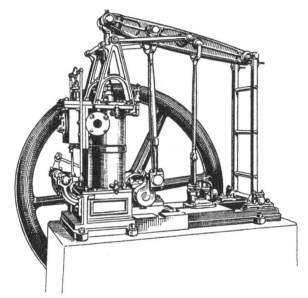

Fig. 20. Grasshopper engine by Easton and Amos, 1862.
In the Science Museum, London.

on four columns or legs; two connecting rods depend from the crosshead to a crankshaft below (see Fig. 21). In a simplified form this engine became a great favourite, since it occupied little space and could be relegated to any odd corner where to-day we should instal an electric motor. It is not surprising that it persisted well into the present century; indeed, one is in existence to-day in working order as a stand-by.

On reflection, it is surprising to find what a hold over engineers' minds the indirect type had when it was known that the direct-acting type had been used as early as 1802 by William Symington (1764–1831) in the paddle tug *Charlotte Dundas* which he engined for service on the Forth and Clyde Canal. Moreover, this engine was horizontal and the compactness of the arrangement would, one would have thought, have invited imitation. The persistence

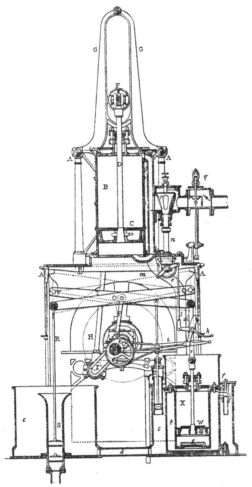

Fig. 21. Maudslay's table engine, 1807.

From Tredgold's *Steam Engine*, 1827.

of the vertical position of the cylinder is perhaps to be explained by the impression held by engineers that in any other position undue wear of the cylinder on one side would result. The fact remains that, apart from the locomotive, the horizontal position was not adopted extensively until about 1825, when Taylor &

Martineau, of City Road, London, built such engines for factory purposes. They were of the type so familiar at one time: the box-girder bed-plate supported the cylinder, with rollers or slide bars for the crosshead, and the crankshaft bearing, with the condenser below. For half a century this type of engine was everywhere to be met with.

One of the early direct-acting engines was the vibrating or oscillating engine. In this arrangement the cylinder is mounted on trunnions and rocks backwards and forwards so that the piston rod can act directly on the crankshaft without the intervention of a connecting rod, thereby saving considerable head room. Murdock, while in Boulton and Watt's employ in Cornwall, made a model of such an engine. Although simple in motion and moving parts, the trunnions were a difficulty, as through one of them the steam must be introduced and consequently a valve gear had to be contrived to distribute it. This Murdock had managed to do with a piston valve—an early instance of its use in the steam engine. The engine did not receive recognition till revived in 1827 by Joseph Maudslay (1801–61), and even then it hung fire until John Penn I (1805–78) took it up and invented a satisfactory valve gear. It became a great favourite for marine work and was to be seen almost universally on river steamboats, where it survived till the end of the nineteenth century. It was never much used on land, so that we are not further concerned with it.

Having now dealt broadly with the different types of engines, something may be said about engine construction and machine details. Before doing so a few words about the development of machine tools is desirable. Nothing strikes one so forcibly when reading about the machine shops of the early part of the nineteenth century as the lack of tools. All that engineers had was, as we have mentioned already, the boring mill and the common lathe. Further equipment was gradually evolved. Henry Maudslay brought out his screw-cutting lathe with slide rest about 1800, but the back-geared lathe for general work was not in use much before 1817. The planing machine is credited to at least three inventors, and its date is not known because they kept it a secret to themselves, but it was divulged about 1820. The machine tool age did not really dawn until Joseph (afterwards Sir Joseph) Whitworth (1803–87) realized that to make accurate machine tools —the field of engineering to which he was then devoting his talents—true screw and plane surfaces had to be originated, for in essence a machine tool can only reproduce what is already embodied in it. Consequently, Whitworth standardized screw threads, originated the surface plate, created plug and ring gauges and made the measuring machine which inaugurated the change to precision engineering that we know to-day, the outcome of which has been to change completely workshop technique.

Coming now to the consideration of constructional details:

Bedplate. An important advance, particularly for the rotative engine, was to support the various parts: cylinder, crankshaft, etc., on a cast iron foundation or bedplate, thus ensuring alignment and doing away with the necessity to support the framing by building it into the walls of the engine house. Murray is usually credited with the introduction of this improvement into practice in an engine built by him in 1802; the utility of the bedplate was such that it was not long before every maker adopted it.

Condenser. Although in 1766 Watt in his condenser used a simple jet of cold water to condense the steam, he had originally intended that the cooling water should not come into direct contact with the steam, and he had entertained the idea of a surface condenser. In 1766 he experimented with tubes and narrow cells made of flat plates closely packed together with cooling water circulating outside. In fact, he did considerable experimental work in this direction because he realized that the air pump discharge could thereby be reduced. Mechanical difficulties were too great at the time and he dropped the idea. It was not until 1834–7 that Samuel Hall (1781–1863) revived the surface condenser, mainly for marine engines, but again it suffered eclipse partly on account of cost, but more because it became clogged with the tallow so freely used as a lubricant in those days. A century elapsed before, with mineral oil lubricant, the surface condenser came to stay.

Packing. The piston of the atmospheric engine was kept tight, as we have seen, with soft packing held down by weights and sealed with a layer of water on top. Watt was precluded from using this seal by reason of his new construction and he had endless trouble until he devised the hemp and tallow packing squeezed down by a junk ring, which will be seen, not too clearly, in Fig. 13. This met low-pressure requirements for over a century. As early as 1778 Watt had inserted blocks of anti-friction metal in soft packing, but it was reserved for the Rev. Edmund Cartwright, D.D. (1743–1823), to employ complete metallic packing and to patent it in 1797. In 1816 John Barton (Patent No. 4062) introduced segmental wedge packing, pressed outwardly by springs, and this was quite extensively used (see Fig. 22). Barton's packing has since been modified

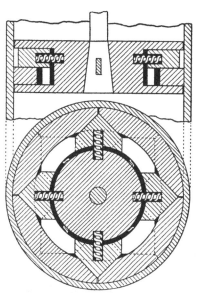

Fig. 22. Barton's metallic piston packing, 1816.

From Tredgold's *Steam Engine*, 1827.

and has been followed by many others, of the same type; in most of them one or more rings, with a saw cut through each, fit the cylinder and are forced radially against the walls and axially against the piston body by flat, volute or helical springs. As regards the piston rod, Watt does not seem to have had any trouble; his stuffing box and gland met all requirements and it has remained with obvious improvements, to be mentioned in a subsequent chapter, to the present day.

Valves and valve gear. To distribute the steam to the atmospheric engine cylinder, a segment of a disc was used actuated by the plug-rod hanging from the beam. Watt employed drop valves similarly actuated for steam, equilibrium and exhaust. In place of these Murdock in 1799 patented (No. 2340) the slide valve—a brilliant invention—as it took the place of the three valves mentioned. Briefly described, it is a box with ports in it caused to slide to and fro in a casing close against the cylinder provided with corresponding ports. The slide by its motion puts the steam and exhaust ports alternately into coincidence. Murdock's slide was circular, or more usually D-shaped, in section, with the flat side next the cylinder; hence it was known as the D-valve—the long-D if it covered the whole valve face, or the short-D if it was in separate portions, one for each port: the exhaust passed out through the central tubular waist. It is to be remarked that in his specification Murdock claimed the oscillation of the valve on its axis as an alternative to sliding lengthwise, but he did not use it in practice; later this kind of valve assumed great importance.

Murray, in his patent of 1801, simplified the slide valve further and eventually it reached the form of a box sliding over three ports on the cylinder face (see Fig. 23). This, the simplest form, is known as the locomotive D-valve from its wide application in the locomotive, but it has been applied in every type of low-pressure engine. It is a property of the valve, not realized at the time it was introduced, that by adding "lap" to it, it will cut off the steam before the end of the stroke and thus secure expansive working. The modifications to which this simple valve has given rise are almost legion.

To give motion to the slide valve, that ingenious mechanism, the eccentric—in reality a crank of small radius—was devised at Soho some time between 1799 and 1800, due it is believed to Murdock and known as the "eccentric circle"; it was not patented, as might have been expected. The eccentric, keyed on the shaft, is encircled by a strap to which a rod is fixed, and this rod actuates the valve. The combination of Murdock's eccentric and Murray's D-valve has received as wide exemplification as all other valve gears put together.

Reversing engines. We have mentioned already that circumstances frequently arise where an engine is required to go first in one direction and then in the other, for example, in winding from mines. The way in which this was first effected was, as we have said above (p. 83), to stop the engine at the right point and start it off in the reverse direction. This can be done if the valve gives steam to the cylinder during the whole of the stroke; if, however, the valve cuts off the steam, other methods have to be used. One of the first of these was to leave the eccentric loose on the shaft, but provided with a stop which came up against one or two corresponding stops on the crank-shaft. The eccentric rod had a notch or "gab" which dropped on and engaged with a lever connected to the valve spindle. To make the engine move in the direction required the rod was lifted up and the valve worked by hand. When the engine got

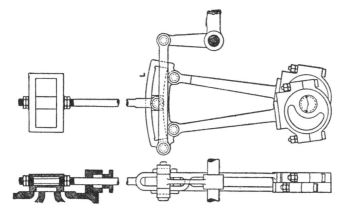

Fig. 23. Howe's link-motion reversing gear, with Murray's locomotive D-valve.

From Ripper's *Steam Engine Theory and Practice.*

going, the rod was dropped back on the gab, the appropriate stops came together and the motion continued. To this succeeded a gear in which there were two fixed eccentrics and two gabs, one for forward and one for backward motion. Later still, these rods were linked together so that one motion of a lever lifted one gab off and brought the other into its place. Thus there was an intermediate position where neither gab was engaged. Then came the crowning invention: the ends of the two rods were joined by a curved link, which engaged but could slide over the end of the valve rod. This was the invention of 1842 of William Howe of Chesterfield; it was at once taken up by Robert Stephenson, who fitted it to a locomotive. Hence it is often spoken of as Stephenson's link motion (see Fig. 23), but more correctly as the shifting link. It has been more widely used and in all types of engines besides locomotives than any other reversing gear. It led to the invention of the stationary link motion of Daniel (afterwards Sir Daniel) Gooch, 1843, and to the straight link motion of Alexander Allan, 1855. Egide Walschaërts of Brussels in 1844 introduced a reversing gear in which only one eccentric was used. This gear did not receive the attention it merited for half a century, but is now the most widely adopted of any for the locomotive.

Up to this point little has been said about boilers, so that we must now pause to give that important part of a steam plant some consideration.

CHAPTER VII

LAND BOILERS UP TO 1850

Early boilers based on brewing copper—Waggon boiler—Flue boilers—
Cylindrical boiler—Woolf's boiler—Elephant boiler—Cornish and Lancashire boilers—Sectional boilers.

THE boiler, though an integral part of every steam
plant, was in the early days so subsidiary to the engine
that it occupied the position of the poor relation; since
then it has steadily advanced in relative importance, *pari passu*
with the rise in steam pressure, till to-day it may be said to be
an equal partner.

As will have been gathered from what has gone before, the
boiler for the atmospheric engine had no pressure to sustain
worth mentioning; it was little more than a tank that had to be
strong enough to resist the pressure due to the weight of water
that it contained. Design, construction and setting were borrowed, there can be little doubt, from the under-fired brewing
or domestic copper; of course, in addition, the boiler had to be
supplied with a lid or cover, usually of lead, for it was necessary
to keep air from getting into the steam, because this materially
slowed down the working of the engine. In shape the boiler
was domed in the upper part, with inclined or flanged sides,
and with a segmental bottom, below which was the fire grate
and ash pit. The boiler was set in brickwork so that the waste
gases passed completely round the exterior before reaching the
chimney, a course known as the "wheel" draught. What the
boiler and its setting were like will be gathered from Pl. II and
from Figs. 7 and 9. From its shape it was known in different
districts as the "haystack", the "beehive", or the "balloon"
boiler; under the latter name it survived in Staffordshire until
late in the nineteenth century; indeed one or two are still left,
in use as water tanks.

Copper was the material first used for the shell, but when

hammered wrought-iron plates became available commercially, that is about 1725, they were substituted, although they were only obtainable in lengths of about 2 ft. Rolled iron plates, usually $\frac{5}{16}$ in. or $\frac{3}{8}$ in. thick, did not become available generally till 1795. Boiler plates were joined by the well-understood method of riveting; the diameter of the rivet, its pitch and the lap of the plates, appear to have been determined by rule of thumb. The cost of a boiler may be judged by the account for one of 12 ft. diameter constructed in 1733 at Jesmond Colliery, Newcastle-upon-Tyne; it worked out thus:[1]

The Charge of the Iron Plates viz. 40 long plates & 40 short ones	£80 . 00 . 00
The Charge making do.	12 . 00 . 00
The Top of the Boyler, Lead abot. 38 Hundrd weight at 18s p. Hundrd weight	34 . 04 . 00
	126 . 04 . 00

Boiler-making was entrusted to the blacksmith, but became a separate though ancillary trade in the Midlands about 1790. Boulton and Watt got their boilers from Thomas Horton of West Bromwich and from John Wilkinson of Bradley and Bersham.

To increase the heating surface—practically the only essential of a boiler at that time—relative to the water content, the next step was to elongate the boiler, retaining the same cross-section, and to supply it with flat or rounded ends, thus arriving at the waggon boiler (cf. Fig. 13, p. 76). It was externally-fired and set with a wheel draught as before. This boiler, provided occasionally with a through flue of approximately square section, remained standard throughout Watt's time. The waggon boiler has been frequently attributed to him, but the evidence is all to the contrary; boiler design was one of the matters to which he did not devote much attention. All he said himself about boilers was that he "somewhat improved the form and adjusted the proportions", and he never claimed to have done more than this. The waggon boiler stood its ground for about half a century

---

[1] *Trans. Newcomen Soc.* XVII, p. 136.

after Watt's time, in spite of the inherent weakness of its shape. This weakness revealed itself as the pressure was gradually raised more and more above the 5 lb. per sq. in. of Watt's day.

Other materials, among which we may mention wood and stone, were employed in the construction of boilers. Many persons doubt whether they ever were used, but that such was the case will be understood when it is stated that it was only the shell that was made of these materials and that the grate and fire were accommodated in an internal firebox or in a through flue, of copper.

With the advent of the high-pressure engines of Trevithick and Evans, a great step in advance in boiler-making was necessarily made. Both engineers realized almost instinctively that a spherical or cylindrical shape was the only proper one. Trevithick, as we have said, at first made use of cast iron and in his patent of 1802 he showed a globular boiler externally heated, but a year or two later he brought out his well-known boiler consisting of a cast-iron cylindrical shell, to the front of which a wrought-iron plate was bolted; to this front was riveted a U-shaped flue or tube containing the furnace (cf. Pl. III). One merit of the boiler was that it could be cleaned properly since the front was readily removable.

Evans arrived at the cylindrical shape about the same time, and he is quite specific as to its advantages. Writing in 1805 he says:[1]

As we mean to work with steam of great elastic power, say 120 pounds to the inch above the atmosphere, it is necessary in the first place to discover true principles...that we know how to construct them with a proportional strength....A circular form is the strongest possible, and the less the diameter of the circle, the greater elastic power it will contain. Therefore we make cylindric boilers not exceeding 3 feet diameter, and to increase their capacity we extend their length to 20 or 30 ft. or more, or increase their number....To save fuel we construct boilers consisting of two cylinders, one inside the other...the space between them contains the fire....This boiler is enclosed in brick work and the flue returned along the under side

[1] *Abortion of the Young Steam Engineers' Guide*, 1805, p. 22.

of the outer cylinder.... These boilers are made of the best iron rolled in large sheets and strongly rivetted together. The ends may be made of soft cast iron, provided the fire or flue be kept from immediate contact with them, as cast iron is liable to crack with the heat. . . .

The first kind of boiler that Evans describes is similar to what was known in England as the egg-ended boiler—a simple cylinder with hemispherical ends, externally fired. The date, 1805, is the earliest we have been able to find for the use of this type of boiler. The second boiler described by Evans is, except in proportions (see p. 94 and Pl. IV), essentially that known in this country as the Cornish boiler, which received its name from that part of the country where it originated. There is not much doubt that Evans made and used this boiler at as early a date as that of this description, whereas the evidence as to the date when the same type was first made in England is inconclusive. We know that such a boiler was in service before 1812, for we have this statement by one of Trevithick's workmen:[1]

About 1812 Captain Trevithick threw out the Boulton and Watt waggon boilers at Dalcoath and put in his own, known as Trevithick's boiler. They were about 30 feet long, 6 feet in diameter with a tube about 3 feet 6 inches in diameter going through the length. There was a space of about 6 inches between the bottom of the tube and the outer casing.

We have also the direct evidence of the boiler constructed by Trevithick early in 1812 to supply steam to a thrashing engine made for Sir Christopher Hawkins of Trewithen, Cornwall. This boiler (see Fig. 24) is preserved in the Science Museum, South Kensington, and it will readily be seen that it is what we know as the Cornish type. There is no direct evidence that Trevithick designed the boiler, but there cannot be any reasonable doubt that he did so; tradition certainly has always attributed it to him. It proved so satisfactory for most duties that it has continued to be made and is indeed not yet entirely extinct; for example, it is to be found in industrial plants for providing process steam.

[1] Trevithick, Francis, *Life*, II, p. 157.

As time went on and still greater heating surface in a given space was called for, the obvious course was to increase the diameter of the shell, and put through it an additional cylindrical flue. This, subsequently known as the Lancashire boiler, was patented in 1844 (No. 10,166) by William (afterwards Sir

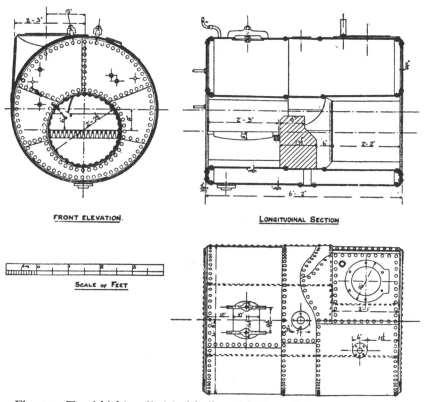

FRONT ELEVATION.

LONGITUDINAL SECTION

SCALE of FEET

Fig. 24. Trevithick's cylindrical boiler with through furnace tube, 1812.
In the Science Museum, London.

William) Fairbairn (1789–1874) and John Hetherington of Manchester (see Fig. 25). Fairbairn stated that[1] "it was originally devised with a view to alternate firing in the two furnaces in order to prevent the formation of smoke" and not as we have said for the purpose of obtaining greater heating surface, although that, of course, followed. The Lancashire boiler may be said to have had, in Great Britain, the longest run so far of any boiler, for

[1] Fairbairn, *Mills and Millwork*, 1861, Pt I, p. 256.

it is still to be met with everywhere. How, by steady improvement, engineers have managed to maintain it in this position will be enlarged upon in a succeeding chapter. It is instructive to be able to compare the cubic contents and space occupied by the boilers we have been discussing in relation to heating surface, thus:[1]

TABLE I. *Tank boilers compared on a basis of 500 sq. ft. heating surface*

Type	Content of water cub. ft.	Space occupied cub. ft.
Haystack	1005	10,000
Egg-ended	775	4,000
Lancashire	404	2,700
Cornish	375	2,900

Concurrently with the development of the tank or shell type of boiler, development along entirely different lines was going on, and this was the construction of a boiler built up of an

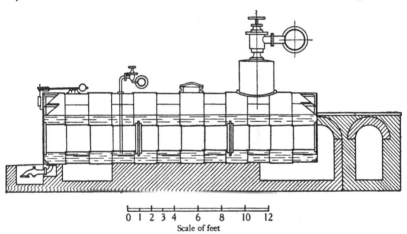

Fig. 25. Fairbairn's Lancashire boiler, *c.* 1845.

aggregation of small elements—spheres, sheet metal cells or tubes—interconnected with each other. Two generic types of such boilers may be discriminated: one in which the water is

---

[1] *Proc. Inst. Mech. Eng.* 1871, p. 235. The data are those of E. B. Marten, based on wide knowledge.

inside the element and the fire outside, now known as the water-tube or tubulous type; the other, in which the water surrounds or drowns the element and the flame traverses the interior, known as the multitubular, fire-tube or locomotive type. The latter type, as one of its names indicates, is almost universal in the locomotive. It was brought into being in an almost fully developed form within a couple of years, 1828–30. Marc Seguin, in France, designed and patented a boiler of this type and applied it to a locomotive on the Lyons and St Étienne Railway in 1828. Independently, in the following year Robert Stephenson and Henry Booth brought out their boiler of the same type for the "Rocket" locomotive, their object being to keep down its weight so that it should comply with the conditions laid down for the locomotive trials, for which it was entered at Rainhill in that year. The "Rocket" duly competed and its success in winning the prize was so marked and due so obviously to the boiler and to the method of inducing the draught by sending

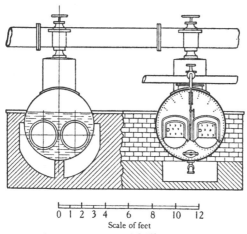

Scale of feet

From his *Mills and Millwork*, 1861.

the exhaust steam as a jet up the chimney, that a new era in locomotive construction and development was thereupon inaugurated. The final step was to incorporate the furnace or firebox, hitherto independent, in the boiler structure, thus making an integral whole; this was effected in the "Northumbrian" locomotive of 1830, and at once the construction became standard in locomotive practice. The boiler is compact, of great evaporative power in proportion to its weight, and offers few difficulties in construction. Its subsequent development, largely one of detail, is outside the scope of this volume. Naturally, the type has been adopted for stationary purposes, namely for portable and semi-portable engines mostly for export to backward countries. Furthermore, the multitubular principle will be found embodied in most vertical boilers; one of the best known in this class is the Cochran boiler, the joint invention of Edward Crompton and J. T. Cochran in 1878, and largely employed in small power installations where other types are inapplicable.

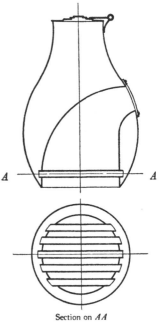

Very different from the case of the multitubular boiler is that of the water-tube type. Its path has been strewn with the discards of almost every conceivable combination of the elements already specified. It becomes necessary, therefore, to trace the chequered history of this type of boiler in some detail and mark as succinctly as possible the protean changes that have taken place in it. The idea of the water tube is very old; as applied to keeping liquids hot we are able to point to a vessel for this purpose in which an internal firebox and water-tube grate

Section on *AA*

Fig. 26. Section of water heater from Pompeii, A.D. 79.

Reproduction in the Science Museum, London.

bars are embodied (see Fig. 26). This vessel was found at Pompeii and consequently dates anterior to the burying of that city by volcanic eruption in A.D. 79.

Conditions are radically changed, however, when internal pressure has to be met, for then a number of difficulties are encountered: rendering tight the multiplicity of joints; the sufficiency or otherwise of soldering, welding or riveting; and allowances for inequalities of expansion. But beyond these are other considerations that must not be, but frequently are, over-looked, such as provision for circulation of the water, ready liberation of the steam formed, and last but not least, provision for cleaning the elements. Each of these desiderata gives rise in turn to a host of conditions nearly always conflicting with one another. No wonder then that practical success in making a water-tube boiler was little more than a dream until the nine-teenth century was well on the way.

Here it may be remarked that practically every idea, every expedient, that has been made the basis of a successful boiler was brought out and tried experimentally during these "dark ages". The records of the Patent Office can be adduced to substantiate this statement, and they can be cited further to invalidate the claims of every fresh inventor. The fact that so much talent has been devoted to the problem is some measure of its difficulty. The truth is that it is not ideas that ever were lacking, but materials of high enough quality, workshop technique and machine tools. To illustrate these points and at the same time to show development chronologically, a few examples must be resuscitated from the scrap-heap.

Possibly the earliest attempt to make a pressure water-tube boiler was that of William Blakey. His boiler, patented in 1776 in the Netherlands (see Fig. 27), had three tubes D, E, F in series, zig-zag wise, inclined in a brick chamber A, with three fire places B, B, B. The bottom tube only was drowned, so that with the other two he must have got a certain amount of super-heat; the germ was there but unrecognized. Circulation must have been defective, but the boiler did not fail because of that, but because the tubes, which were of copper brazed at the seams, would not sustain the pressure demanded.

James Rumsey (1743–92), an American, came to England and patented in 1788 several designs of boilers. One had a firebox with flat top and sides with horizontal tubes across the firebox connecting the water spaces. Resembling this was another which had a cylindrical firebox, in which was a coiled tube whose ends were connected to the water spaces. We have here the germ of the "coil" boiler. Another of his designs was the vertical tubular boiler resembling the donkey boiler still made to-day. The drawings in his patent specification are too crude to be worth reproducing. Rumsey's death at the early age of forty-nine cut short these promising inventions.

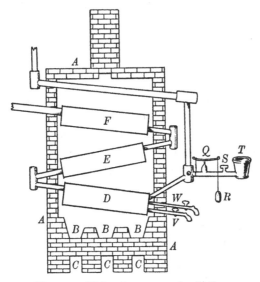

Fig. 27. Blakey's water-tube boiler.
From his Netherlands patent specification, 1776.

Arthur Woolf in 1803 (Patent No. 2726) successfully intro-duced a multiple large tube boiler made wholly of longitudinal cast-iron pipes in tiers, united by downcomers with bolted flanges (see Fig. 28).[1] He made provision for expansion but nevertheless it asserted itself; this caused leakage and breakage of the downcomer necks and proved to be a difficulty that the inventor never entirely overcame, so that by 1824 he had abandoned the boiler. It had one great advantage, and that was freedom from liability to explosion; this is shown by Woolf's evidence before a Select Committee of the House of Commons in 1817, when he stated that he had never had an accident with it since its introduction. Woolf supplied these boilers for his compound engines, which, as mentioned previously (see p. 100), he exported to the Continent. His quondam partner, Humphrey

[1] Taken from *Phil. Mag.* 1805, XVII.

Edwards, who went to Paris and built Woolf engines there, took the boiler along with him.[1] He modified it in 1815, still employing cast iron as his material, till it consisted of two small drums over the fire, and superposed above them a large combined drum and steam receiver. In 1825 he replaced cast iron by wrought iron, still adhering to the same design, and thus surmounted previous difficulties. A standard type was thereby arrived at, and this was brought into this country about 1850, as something new under the name of the "French" or "elephant"

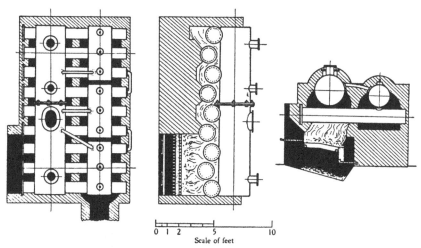

Fig. 28. Woolf's cast-iron sectional boiler, 1803.

From *Phil. Mag.* xvii.

boiler. It achieved some measure of success for textile mills, but by that time the Lancashire type had got a hold upon mill-owners—an exceedingly conservative race—so that the "elephant" boiler did not manage to make its way to any considerable extent. On the Continent, however, where the Lancashire boiler never penetrated, the "elephant" boiler continued to do yeoman service until well on in the present century; it has now been replaced by the small water-tube type.

In 1804 Col. John Stevens of Hoboken, N.J., made a "porcupine" boiler, in which the heating surface was in the form of small tubes projecting from a main drum. His son patented it in

[1] *Trans. Newcomen Soc.* xiii, Pl. IV.

England in 1805 (No. 2855), and fitted this boiler in a steam launch to supply an engine driving a screw propeller. It did not lead to further progress, however, and remained a surprising anticipation of subsequent development. This boiler and engine are preserved in the National Museum, Washington, D.C. The thimble-tube boiler of Thomas Clarkson (1864–1933) is a modern exemplification of this type.

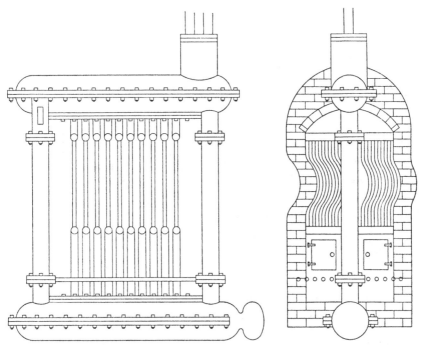

Fig. 29. Eve's water-tube boiler.

From his patent specification, 1825.

In some of these boilers, as in many to be mentioned later, the necessity for providing for circulation of the water does not seem to have been realized. Joseph Eve of Augusta, Georgia, U.S.A., seems to have been the first to provide for this by having external downcomers, as shown in his English patent of 1825 (No. 5297). The boiler (see Fig. 29) does not appear to have got to the practical stage, however.

The introduction of steam carriages on common roads in the 'thirties of last century created a demand for quick-steaming boilers. This demand produced from inventors a crop of small tube and sectional boilers, for it will be realized that this was anterior to the general adoption of the locomotive type of boiler. We can mention a few only of the more original.

Sir Goldsworthy Gurney (1793–1875) in 1826 made a boiler of wrought-iron pipes bent to a hair-pin or U shape; the ends

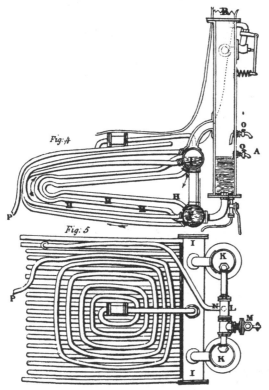

Fig. 30. Gurney's water-tube boiler, 1827.
From Gordon's *Elemental Locomotion*, 1834.

were screwed into horizontal cast-iron headers and held there by check nuts. These headers were joined by downcomers, and were connected by bends to vertical steam separators of wrought iron. The lower legs of the pipes constituted the grate bars, and brick baffles were inserted to direct the gases among the tubes (see Fig. 30). Those were the days of butt-welded tubes and their almost ineradicable defect of bursting asserted itself; besides it was admitted that the circulation of the water was defective. It was tried as a land boiler, but the eclipse of its use in the steam carriage militated against further development, so that nothing further came of it.

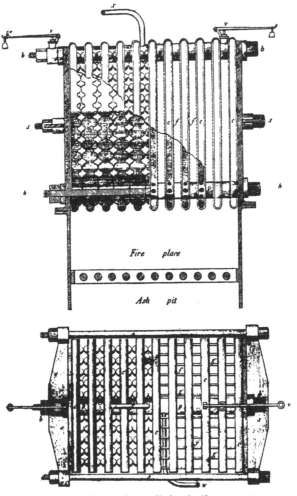

Fig. 31. Hancock's cellular boiler, 1827.

From his *Narrative of Twelve Years' Experiments*, 1838.

Walter Hancock (1799–1852), the most successful of the steam-carriage pioneers, in 1827 brought out a boiler (see Fig. 31) that consisted of narrow flat chambers or cells with projecting humps, corrugations or distance pieces to keep the chambers sufficiently far apart for the hot gases to pass between them; the cells were held together by tie rods. Our present-day hot-water radiators to some extent recall what this boiler was like. It will be observed that there was no provision for cleaning the insides of the cells.

Another boiler of some merit designed for the steam carriage was that of William Altoft Summers and Nathaniel Ogle of 1830; it embodied the novel idea of fire-tubes inside and con-centric with water-tubes. The latter were joined vertically to horizontal boxes through which the fire-tubes passed and these were held in place by nuts, which served to draw the joints tight. Another steam-carriage boiler resembling this was that of Dr William H. Church, 1832. This type of boiler has lately received exemplification in spite of difficulties of manufacture.

The development of the steam carriage was arrested in 1835 by legislation, and by the imposition of prohibitive tolls, pro-moted by the interests in Parliament of the horse-coaching fraternity and the turnpike trusts. Thus this hopeful boiler development was nipped in the bud and the spurt was over. However, the need for improved transport had become insistent, and the effect of baulking the steam carriage was not to help the horse coach, but rather to sign its death warrant, by giving a fillip to the spread of railways. These in their turn led to the rapid development in a few years time of the locomotive and the embodiment in it of the multitubular boiler already mentioned.

We could go on at length to describe other tube boilers— the patent records of this and other countries are full of them— all that need be said, however, is that nearly every one suffered from one or more inherent defects, such as have been mentioned; construction demanding more than the workshop capabilities of the time; materials of insufficiently high quality; differential or unequal expansion causing leakage, etc.; defective circulation; and difficulty of access for cleaning.

Combustion. Efforts to improve combustion by grading the coal, multiplying the spaces between firebars without increasing the total area for air admission, or by forming the grate in a series of steps have been effective. Most important has been the introduction of the mechanical stoker originating in the attempt to prevent the emission of black smoke, the result of burning bituminous coal. Stokers are of two kinds—coking and sprinkling; the latter can be further subdivided into the underfeed and overfeed types. The coking stoker owes its origin to William Brunton who patented it in 1819 (No. 4387); this had a flat revolving grate. In 1822 he took out a further patent (No. 4685) for a straight-line moving stoker which he termed the "peristaltic" grate, owing to its action resembling that of the muscles in the alimentary canal; he caused each alternate grate bar to rise slightly, travel forward with the fuel, fall back to the grate level and return to its original position, when the remaining bars went through the same movements.

In 1822 John Stanley, a Manchester blacksmith, patented (No. 4692) his "fire feeder"—an overfeed stoker in which the coal was sprinkled over the grate at frequent intervals. In 1834 John George Bodmer patented (No. 6616) the travelling-grate stoker, subsequently developed by John Juckes (Patent No. 9067 of 1841) and known as the chain-grate stoker, the most widely used of all types. It is sad to have to say that Juckes died "almost a beggar in the streets", a fate too often shared by inventors. The first to devise a really practical underfeed stoker was Richard Holme (Patent No. 6503 of 1833); he used a tapered screw working in a trough.

Feed-water heating. To heat the water supplied to the boiler was from the first recognized as a desideratum. Savery, as we have seen, installed a supplementary boiler at the side of the main one. Newcomen and Watt took their feed water from the eduction pipe and the hot well respectively. Trevithick, realizing the loss of heat that resulted in his engine by dispensing with the condenser, passed the exhaust steam through a pipe concentric with a larger one; through the annular space the feed water contraflowed.

The fundamental idea of utilizing some of the heat in the waste gases from the boiler—gases which may be at a temperature of 600° F. and over—occurred to many minds. An apparatus to be effective must be close to the boiler in the flue leading to the chimney. Passing over tentative experiments, we come to the design of Edward Green (1799–1865) of Wakefield of what came to be known as the fuel economizer. It consisted of a stack of cast-iron pipes connected in parallel at top and bottom to semi-cylindrical chambers. The cold feed was introduced into the bottom chamber and the hot feed taken off from the top. In the first apparatus all went well till gradually the pipes got sooted over; this not only constricted the flue area and impaired the draught but, as soot is an excellent non-conductor, left the feed water cold. Attempts to keep the pipes clean by hand

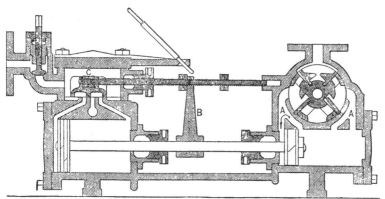

Fig. 32. Worthington and Baker's direct-acting feed pump.
From their patent specification, 1849.

proving futile, Green designed balanced mechanical scrapers, automatically worked by a small engine; this arrangement he patented in 1845 (No. 10,986). Numerous as have been the modifications of the apparatus in detail, the principles of parallel upward flow of the feed water and mechanical scraping of the pipes have been preserved, and the apparatus is still standard.

Feed apparatus. The atmospheric engine boiler was fed, as we have seen, by a stand pipe in the latter. When the height of this (2 ft. = 1 lb. pressure roughly) became unwieldly, a simple plunger pump worked off the beam or from the engine shaft was substituted. As installations grew larger, e.g. boilers in a range, an independent "donkey" pump became usual. The first to supply this need and the one whose name is closely associated with the independent pump is Henry Rossiter Worthington (1817–80) of New York, who patented on September 7, 1841, his non-rotative pump with the steam and the pump cylinder directly in line. In 1844 he patented his spring-thrown valve and later his steam-thrown valve. In pumps such as these, i.e. without a flywheel, there is a possibility of the valve sticking and the pump stopping. To obviate this, Worthington and his partner, William H. Baker, patented the "relief valve" motion. The steam valve was moved by tappets at the end of each stroke and the necessary "kick" was administered by removing the load suddenly at the end of the stroke; this was effected by making the piston uncover by-passes which placed both ends of the cylinder in communication. The pump as illustrated in this patent is shown in Fig. 32; *B* is the tappet and *AA* the by-passes. Subsequently to 1850 Worthington made further improvements and his lead has been followed everywhere.

Here we must pause awhile and go back to the engine, to resume in a subsequent chapter the consideration of the progress of the boiler and its details.

CHAPTER VIII

FROM HEYDAY TO RECESSION,
1851 TO THE PRESENT DAY

Obsolescence of the beam engine—Refinement of the horizontal engine—
Corliss engine—Porter-Allen engine—Inverted or steam hammer type—
High-speed engines for electric lighting—Threat of the gas engine—
Bramwell's prophecy—Large units for power stations—Uniflow engine.

THE industrial progress of the world in general and of
Great Britain in particular in the half century preceding
the date when this chapter opens had been phenomenal.
Nothing marked so concisely the vast strides that had been
made than the first large-scale Industrial Exhibition—the great
Exhibition of all Nations of 1851—held in London. The display
of raw materials, manufactured products, machinery, tools and
engines then brought together from the ends of the earth was a
revelation to everyone; so busy had industrialists been in their
own small spheres that they had not had time earlier to pause to
take stock of progress generally. Incidentally, the influence that
this and succeeding exhibitions has had in stimulating the advance
of industry can hardly be overestimated. Capital inventions
had been poured forth almost like a flood during the preceding
century, and their effects were now manifest. The burden of
human toil was beginning to be lightened and the standard of
living was being raised. That many drawbacks accompanied the
advance was recognized, but they were believed to be merely
temporary, and bound to disappear with further invention and
further progress. Everything seemed to be for the best in the
best of all possible worlds. An age of universal peace and
prosperity seemed about to be inaugurated.

In this atmosphere of optimism and hopefulness, engineers
found themselves. No limit could be set to the opportunities
that were being presented. Great Britain, if it was not already
the workshop of the world, bid fair to become so. Machinery
was needed for every industry, railways were spreading their

networks over most countries, iron shipbuilding had been established and was helping to enlarge greatly overseas transport.
Small individual workshops were growing into large establishments. Technical capacity had kept pace with the advance; new
materials had been introduced or old ones improved; command
over them had increased, e.g. by the steam hammer, the rolling
mill and the hydraulic press. Machine tools for every duty had
been developed; precision of workmanship had increased.

As far as the steam engine was concerned, these advances
exerted a great influence upon its construction, but apart from
that, its design was undergoing far-reaching changes. The beam
type was disappearing, and its place was being taken by the
slow-running horizontal engine. This again was improved in
economy by increase in steam pressure bringing in compounding,
and by increased speed. Great attention was consequently devoted
to valve gear. The high-speed engine with forced lubrication was
developed to meet the demands arising from the introduction
of dynamo-electric machinery for lighting and later for power
purposes.

The greatest change amounting to a revolution during the
period under review was caused by the advent of the steam
turbine. If to this we add the development of the internal
combustion engine, we have two factors that have been instru-
mental in ousting the reciprocating steam engine from its former
proud position and relegating it to comparative obscurity, so
that, broadly speaking, it may be said that its day is over.

We must now consider in some detail the course of development during the period under consideration. The beam engine
still held sway although challenged from all sides. It survived
longest to supply towns with water. The Waterworks Clauses
Act, 1847, gave an impetus to "constant supply" distribution
of water at high pressure which we enjoy to-day almost universally. This necessitated engines which could respond readily
to sudden variations in pressure and consequently in speed.
Hence it was that the Cornish engine, so efficient under steady
load, gave place to the high-pressure compound rotative engine
with a flywheel, which was flexible, safe and easy to control.

Messrs Simpson & Co., of Westminster, built such engines from 1848 onwards, and the type became a standard one in England. In the United States similar engines based on river steam-boat practice—the well-known "walking beam" engines—were introduced, and many different varieties were tried; notably among them should be mentioned the engine designed and constructed by E. D. Leavitt, Junr. for Lynn Waterworks, Mass. It should be realized that these varieties were essentially the mill engines of Watt. Like them they were provided with drop valves for steam and exhaust, and it was in this direction that development took place. Frederick E. Sickels, an American engineer, had patented in 1841 a drop valve cut-off gear, and had used it on the engine of the river steam-boat "South America". Subsequently, he invented a means whereby the admission valve could be detached at any desired point of the stroke so as to vary the expansion.

About this time it was realized that a governor could be used to greater effect than had been done by Watt. To obviate over-much external resistance, which militates against sensitiveness, the valve to be moved by the governor was balanced. The governor could then be used to vary the cut-off or alternatively it could be used to operate a relay which itself does the actual work. This is frequently found in steam turbines where it is associated with an emergency or runaway valve which comes into operation should an accident occur.

The first to attach a governor to a drop cut-off valve motion was George Henry Corliss (1817–88), and not only so, but he introduced the cylindrical rocking valves, for which he took out his well-known patent on March 10, 1849. This patent was reissued on July 29, 1851, for 15 years, and he continued to patent improvements on the engine during the rest of his life. Corliss was not brought up to, or even connected with, engineering, but he had a flair for it, and this led him to enter into partnership with two other men in 1844, in establishing an engine works at Providence, R.I. Two years later he began his improvements with the object of decreasing fuel consumption. Eventually, he became the best known engine maker in America, and possibly his high-water

mark was reached with the enormous beam engine that he built
for driving the machinery at the Centennial Exhibition at
Philadelphia in 1876 (see Fig. 33). This engine had two cylinders
40 in. diameter by 10 ft. stroke, the flywheel was 30 ft. diameter
and the speed 360 r.p.m. Before this, however, his engine had
appeared in Europe, notably at the Paris Exposition of 1867,
where it attracted much attention. Its fame spread abroad, and
it was taken up by European builders. In Great Britain it
underwent redesign and has been built almost exclusively in
the horizontal form. The valve gear as applied to a rolling mill
engine 40 in. diameter by 5 ft. stroke built for John Ramsbottom,
Chief Mechanical Engineer of the London & North Western
Railway, at Crewe Locomotive Works in 1866, is illustrated
in Fig. 34. The arrangement of the gear is that patented in 1863
by William Inglis. The valves are cylindrical in section and are
brought tangential to the cylinder wall, thereby minimizing
clearance space. The steam in this space expands but it does not
add to the effective pressure. The steam valves, one to each end
of the cylinder, usually on the top of it as shown, are rocked to
and fro by an eccentric; the two exhaust valves are similarly
but independently operated. The steam valves, however, are not
positively connected with the eccentric, but are moved by it so
far, and then tripped; thus released, they close almost instan-
taneously under the action of springs. Such rapid closing has
been recognized as a desideratum since the days of Newcomen,
whose "tumbling-bob" on the steam cock served that purpose.
The way in which the tripping is effected is by a wrist plate to
which are attached links, one to each valve, provided with
hardened steel plates which engage similar plates on levers
connected with cranks on the spindles of the steam valves. One
plate is caused to slip over the other by the governor at the
predetermined point. The spring is enclosed in an air dash-pot,
by which the quick-closing motion is brought to rest without
shock. It will be realized that this necessitates close governing,
and much ingenuity has been expended in meeting this require-
ment.

In Great Britain slide-valve distribution of steam to the

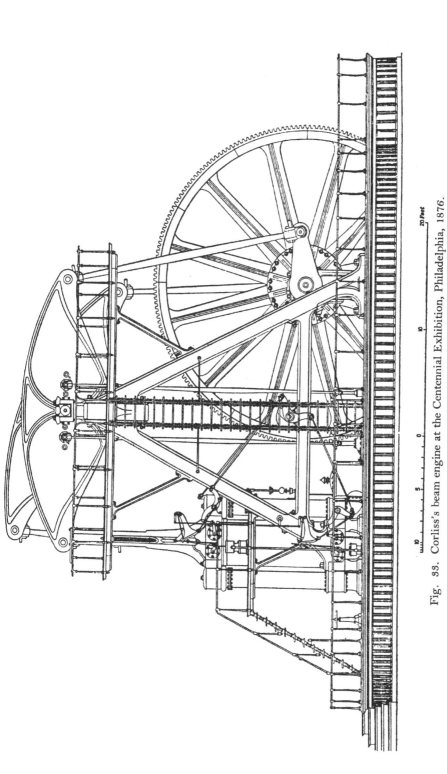

20 Feet

Fig. 88. Corliss's beam engine at the Centennial Exhibition, Philadelphia, 1876.

From *Engineering*, May 19, 1876.

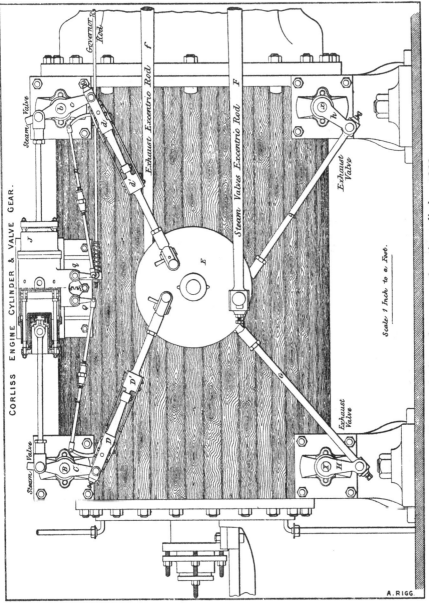

Fig. 34. Corliss valve gear and engine cylinder, 1866.

From Rigg's *Steam Engine*, 1878.

cylinder has been most favoured. As pressure increased and the friction of the valve on its face became excessive, resort was made to the expedient of a relief frame on the back of the valve, or of replacement by a piston valve. A valuable property of a sliding valve is that its movement may be continued so as to control other passages and so combine the duties of several valves.

One of the best ways in which to alter the fraction of the stroke during which steam is admitted is to add another and adjustable valve on the back of the first, and to drive it by a separate eccentric, as in the valve gear of Jean Jacques Meyer (1804–77) brought out in 1842. To make the action automatic, the adjustment of the second valve is placed under the control of the governor. An early contrivance of this kind was that of Zachariah Allen, in 1834. Quite a number of expansion slide valves have been brought out, e.g. by A. K. Rider (1868), by W. Hartnell (1876), by Paxman and others, while of the drop-valve type the best known is that of Dr W. R. Proell, 1881.

During the period we are considering, the horizontal type of engine continued to be a prime favourite for waterworks pumping. We must mention a clever adaptation of the pump by Worthington, already mentioned, and patented by him in 1859 for this purpose. He used side by side two lines of horizontal pumps and steam cylinders; the valve of one engine was controlled by the motion of the piston rod of the other, hence it is known as the "duplex" type. Thus the speed accommodates itself automatically to the load. The engine itself was compounded, and with the addition of compensators has had a considerable vogue. This method of actuating valves has been a fruitful one in other applications.

The horizontal engine was frequently compounded by placing a high-pressure cylinder tandem with the low-pressure one on a common piston rod. If more power was required two such units were placed side by side with the flywheel between. It was slow-running, and an obvious way to obtain more power was to increase the speed. The impetus in this direction arose in the United States from the labours of Charles T. Porter (1826–1910). Like Corliss he was not brought up to engineering, but exchanged the law for it when he became of age. The first job on which he was engaged—the construction of a steam sawing machine—showed defective work by reason of the fluctuations in the speed of the engine. The Watt governor was still the accepted means of control, and it will be realized that it is change in speed of the engine that brings the governor into action; the aim is to make this departure from the normal as small as possible with the least possible time lag or "hunting". To increase this sensitiveness Porter, in 1858, loaded the governor by a weight on the centre sleeve, which tended to keep down the balls without increasing centrifugal action. Porter's governor permitted an increase in the normal speed, and his colleague, John T. Allen, designed a high-speed engine embodying this governor and other departures from current practice: shorter stroke—about twice the diameter of the cylinder; small fly-wheels, and reduction of weight of reciprocating parts. Allen had four balanced valves to the cylinder, two admission and two exhaust, driven by the valve rod through the intermediary of a curved link interposed between it and the eccentric; the position of the valve rod in the link was adjustable by the governor and this controlled the cut-off (see Fig. 35). One of these engines was reported on by a Committee of the American Institute; it was non-condensing, having a cylinder 16 in. diameter, 30 in. stroke, supplied with steam at 75 lb. pressure, making 125 r.p.m. It indicated 125 h.p. with $25\frac{3}{4}$ lb. of steam, or 2·87 lb. of coal per indicated h.p. per hour. A good performance! The Allen engine was exhibited at the International Exhibition of 1862 in London and excited not only astonishment but also mistrust amongst the old stagers, for its speed was about double what was then usual. Nevertheless, it blazed the way and influenced a speed-up generally. British firms, such as Messrs Tangye of Birmingham, and Messrs Robey of Lincoln, followed with improved designs, e.g. trunk guides for the crosshead, which thus assumed a cylindrical cross section; while the trunk itself served as a girder joining the cylinder to the crankshaft pedestal.

Subsequent improvements in governors consisted, briefly in springs to supply the closing power and balls deviating but

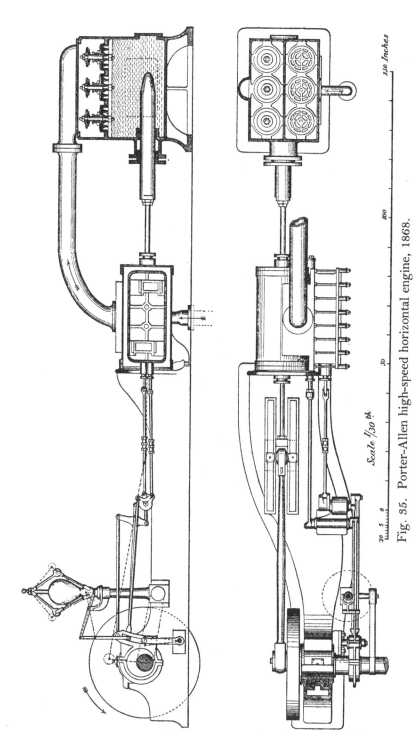

Fig. 35. Porter-Allen high-speed horizontal engine, 1868.

From *Proc. Inst. Mech. Eng.* 1868.

Scale ¹/₃₀ th.

slightly from a horizontal plane to eliminate the effect of gravity, as exemplified in the Pickering, 1862 (see Fig. 36), the Buss, 1870, the Hartnell, 1876, the Hartung, 1893, and the Jahn, 1912, governors. The shaft governor forms another class; this is embodied in the eccentric, whose angular position on the shaft is thereby adjusted so as to alter the advance and travel of the valve.

An engine that had its origin in the steam hammer invented by James Nasmyth (1808–90) in 1839 was known by that name

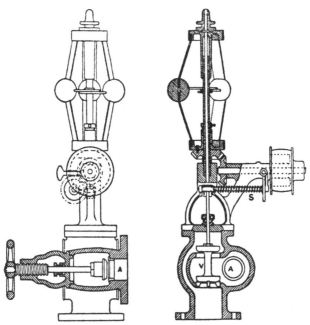

Fig. 36. Pickering's high-speed engine governor, 1862.

From Ripper's *Steam Engine Theory and Practice.*

or as the "inverted" type, because it was vertical with the cylinder above and the crankshaft below where the anvil would be. Nasmyth exhibited one of these engines at the Exhibition of 1851, where its merits were recognized by the award of a medal. The type has been adapted widely for pumps for water, and for air for blast furnaces, but most widely of all in marine practice, where it maintained a dominant position from 1855 till the end of the reciprocating engine period.

A highly original type of steam engine is the three-cylinder high-speed engine with which the name of Peter Brotherhood (1838–1902) is closely identified. He patented it in 1871 (No. 648) and in 1872 exhibited his first engine at the Agricultural Hall, where it excited great interest and was brought rapidly into use (see Fig. 37). The cylinders are fixed radially at 120° to one another in the vertical plane; the crankshaft is overhung and counterbalanced; the pistons are fitted with metallic packing and there are no piston rods because the connecting rods have spherical small ends fitting into the pistons and their other ends embrace a common crank pin. This is possible because the cylinders are single-acting; thus there is no reversal of stress and consequently no knock in the brasses. Naturally this conduces to quiet working at high speeds. The steam is distributed by balanced piston valves actuated from the crank pin. The governor is of the spring-loaded shaft type already alluded to. The crank chamber is enclosed to retain the splash lubricant and to keep out dirt. Piston speeds up to 500 ft. per min. are attained, thus affording large power in a confined space. The engine is made in many sizes, from $2\frac{1}{2}$ in. diameter by 2 in. stroke and 1000 r.p.m., to 7 in. diameter by 6 in. stroke and 500 r.p.m., exerting from $1\frac{1}{4}$ to 55 brake h.p. The engine has found many applications, e.g. as an air motor for torpedoes; it has also been adapted as a hydraulic motor and conversely as an air compressor.

About 1880 the steam engine received an unpleasant jar by the emergence of a rival—the gas engine—which had been steadily gaining ground since its commercial introduction by Niklaus August Otto (1832–91) in 1876. It was easy for engineers to see that to develop the heat within the working substance itself, as was the case in the gas engine, offered a great advantage over the transmission of the heat by conduction through a containing shell, as in the steam engine. Engineers began to look upon the steam engine as out of date and the gas engine as the hope of the future. Lord Armstrong, in his address to the Mechanical Science section of the British Association in 1881, aptly criticized the steam engine as then known thus: "In expanding the steam we quickly arrive at a point at which the reduced pressure on the piston is so little in excess of the friction of the machine as to render the steam not worth retaining, and at this point we reject it...we take the cream off the bowl and throw away the milk." Sir Frederick Bramwell, then Pontifex Maximus of the engineering world in the sense that he spoke with almost papal authority, declared at the same meeting that however much the British Association might then "contemplate with regret even the most distant prospect of the steam engine becoming a thing of the past, I very much doubt whether those

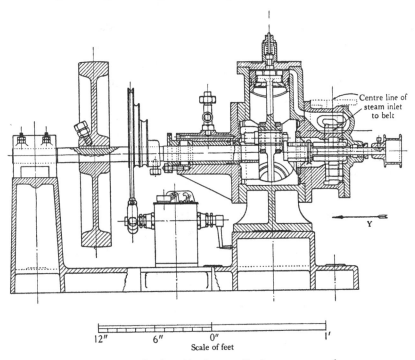

Fig. 37. Brotherhood's three-cylinder steam engine

who meet here fifty years hence will then speak of that motor, except in the character of a curiosity to be found in a Museum". Quite a number of eminent scientific men shared his opinion. He repeated the prophecy in 1888, and a few months before his death he offered a sum to the Association to be devoted to the preparation of a paper by a distinguished man to deal with the whole question of prime movers when 1931 arrived. This address was duly delivered by Sir Alfred Ewing. However, in the interval a greater than the internal combustion engine had arisen in the steam turbine, with which we shall deal shortly, and Sir Alfred had to say: "Steam is neither dead nor dying. On the contrary, its use has immensely developed both on land and sea. To-day it is a much more efficient medium than it was for the conversion of heat into work, and you will find it actuating engines of vastly greater individual and aggregate power than were even imagined when Bramwell spoke."

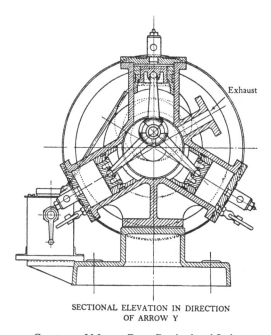

SECTIONAL ELEVATION IN DIRECTION
OF ARROW Y

Courtesy of Messrs Peter Brotherhood Ltd.

After this digression, which shows how true it is that you never can tell, we must revert to the ferment that was now stirring to its very depths the complacency of the older engineers. In 1881, at the very time Bramwell made his pronouncement, electric lighting was arriving on the scene and there was a rush to adapt the existing slow-running horizontal engine to drive the dynamo by using belting to multiply the speed. In vain, governing was not close enough to avoid flickering of the arc lamps, and besides, such plant took up much space where it could ill be spared. What was wanted was a quick-revolution engine that could be coupled up direct to the dynamo. Brotherhood stepped into the breach with his three-cylinder engine, and many other types were invented or introduced. Requirements were more exacting than anything previously demanded, neces-

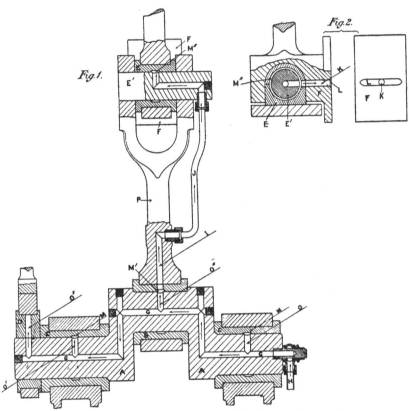

Fig. 38. Pain's system of forced lubrication for engines.
From his patent specification, 1890.

sitating attention to close governing, improved lubrication, balance of reciprocating and rotating parts, and the use of improved materials to reduce weight. In this country nearly all the engines brought out to meet the new need were of the inverted double-acting type, frequently tandem compound. One of the most successful was the engine introduced by George Edward Belliss (1838–1909) of Birmingham, who had had twenty years' experience in marine engine work. The success of this engine was largely attributable to the fact that it embodied, and was the first to do so, the present system of forced lubrication to all parts; this was patented in 1890 (No. 7397) and 1892 (No. 11,432) (see Fig. 38) by Albert Charles Pain (1856–1929), a draughtsman in Belliss's works, and it has since received world-wide application in engines of all kinds from that of a motor car to those of a battleship. The first Belliss engine was built in 1890, at first for driving part of their works, but latterly for supplying the lighting, and continued in use till 1919; it is still preserved in working order because of its interest. The engine is of 20 h.p. and its speed 625 r.p.m. The journals and bearing parts show little signs of wear after twenty-nine years of service during which the crankshaft must have made more than 4000 million revolutions. The most recent engine of this type is compound, shown in Fig. 39. Only one eccentric and valve rod is used to drive both high-pressure and low-pressure piston valves. An oscillating pump worked from the eccentric supplies oil at a pressure of 10 to 30 lb. per sq. in. The lubricating system can be picked out fairly easily.

Many other makers entered the field with engines of similar type. We may mention Allen of Bedford, Bumsted & Chandler of Hednesford, Tangye of Birmingham and Westinghouse of Schenectady. There was another vertical type of entirely different and original design, and that was the engine of Peter William Willans (1851–92) patented from 1884 onwards. The cylinders were arranged tandem with the high-pressure one above, and to eliminate reversals of stress they were single-acting; the guide or crosshead worked in a trunk that acted as an air cushion. Two or else three lines of cylinders acted on a two- or a three-throw crank. The originality was in the steam distribution, which was effected by a long piston valve working inside the trunk and concentric with it. The cylinder itself acted as the common piston rod. Thus the valve travelled with the piston and received its motion relative to the latter from an eccentric on the crank pin. The crank chamber was enclosed to retain the bath of lubricant. The speed was regulated by a spring-loaded

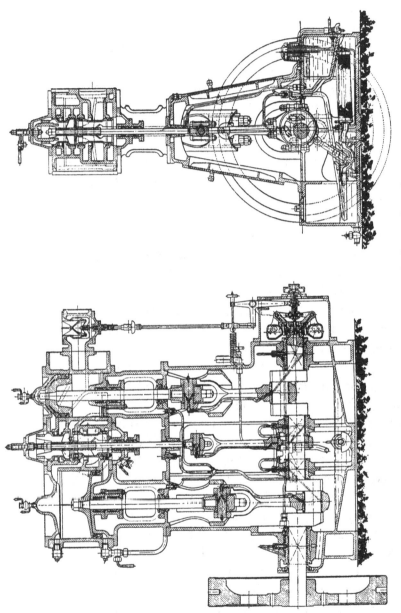

Fig. 39. Belliss's self-lubricating high-speed engine.
Courtesy of Messrs Belliss & Morcom Ltd.

centrifugal shaft governor. These engines were made in large numbers and up to about 500 h.p.; they were in high favour not only for electric lighting, but also as auxiliary machinery on shipboard, etc. Owing to the non-reversal of pressure on the

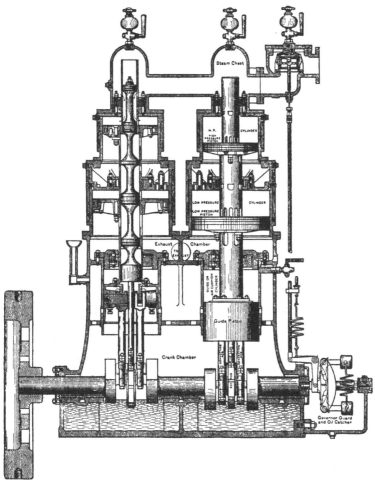

Fig. 40. Willans's central-valve high-speed engine, 1884.
From Ripper's *Steam Engine Theory and Practice.*

crank pin, the engine was unusually silent in running (see Fig. 40).

In the United States the horizontal high-speed engine was favoured, and the best known was that introduced by P. Arming-

ton and Winfield Scott Sims in 1888. This had piston valves and
a shaft governor. In the attempt to develop the utmost amount
of power in a given space, the vertical-horizontal engine was
developed from the latter; this has a horizontal low-pressure
and a vertical high-pressure cylinder both acting on the same
crank pin. This megatherium of the engine world was installed
in New York and later in the London County Council Tramways
Power Station at Greenwich.

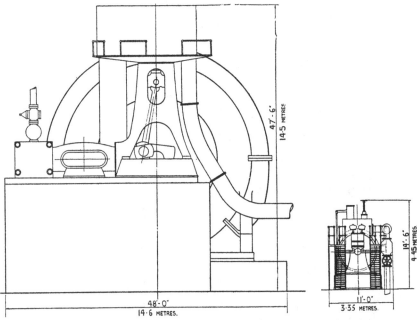

Fig. 41. Space occupied by reciprocating engine and by turbo-alternator
of 3500 kW. output compared, 1906.

Courtesy of Messrs C. A. Parsons & Co.

The enormous size of these engines and the cathedral-like
proportions of the station buildings necessary to accommodate
them can be judged from Pl. V. The high-pressure cylinder
was 33½ in. diameter and the low-pressure 66 in. diameter; the
stroke was 4 ft. This was duplicated, with the alternator between. Steam was supplied at 180 lb. pressure and the engines
ran at 94 r.p.m. The alternator, which was of the revolving
field type, acted as a flywheel and generated three-phase current

PLATE V. VERTICAL-HORIZONTAL ENGINES AT GREENWICH POWER STATION, 1904

Courtesy of E. W. Dickinson Esq.

at 6600 volts and 25 cycles per second. The normal output was 3500 kW. and the overload 4375 kW. The engines were made by John Musgrave & Sons Ltd. of Bolton, Lancashire. The Station which was regarded as the high-water mark of design and construction with reciprocating engines was opened on May 26, 1906. Even then the decision to employ such engines had been called in question by some engineers. It was not long before it was realized that the engines were obsolete and by 1922 every one of the four sets installed had been scrapped and replaced by turbo-alternators. The open spaces left now that the engines are gone—and the same remark applies to other stations—bear mute evidence to the supersession of the recipro-cator by the turbine. The saving in cubic space, particularly in height in buildings, and consequently in capital cost due to one as compared with the other is strikingly shown by the diagram, Fig. 41.

Uniflow engine. For twenty years after 1900, compound and triple-expansion engines continued to be made practically without change for textile mills, for steel works, etc., until the advent of the uniflow engine. As its name implies, the steam flows only one way in the cylinder, i.e. from the hot inlet ends to the cool exhaust middle. The piston acts as the exhaust valve and therefore requires to be equal in length to the stroke, less the clearance space, which is usually about 10 per cent of the stroke. The history of the engine is a long one. Jacques de Montgolfier sketched such an engine in 1825, and Jacob Perkins, in 1827, patented an engine which embodied the principle. Other inventors entertained the idea. More recently, Leonard Jennett Todd, in 1885 (Patent No. 7801), clearly specified the invention, of which he stated the object to be "to produce a double-acting steam engine which shall work more efficiently, which shall produce and maintain within itself an improved gradation of temperature extending from each of the two Hot Inlets to its common central Cold Outlet, which shall cause less condensation of the entering steam, and which shall work with greater economy than has hitherto been the case". The patentee called his cylinder a "Terminal-exhaust" cylinder and the arrangement is shown

in Fig. 42. The "non-exhausting" main slide valves are shown
next the cylinder face; expansion valves are shown on the back
of these; his "piston cut-off valves regulated by the governor"
are on the back of these again, while the "Cold Outlet" is through
the ports at the centre of the cylinder.

While the advantages of the arrangement were realized at
the time, the difficulties involved in construction, to be referred to

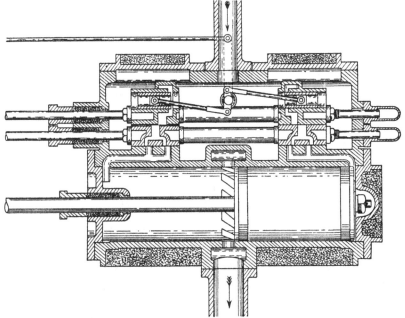

Fig. 42. Todd's "Terminal-exhaust" cylinder.

From his patent specification, 1885.

below, were not overcome, and the invention lay dormant till
1908, when Dr Johann Stumpf of Charlottenburg directed his
attention to the engine; he designed and patented a valve gear
to suit the engine. It was at this time that the name "uniflow"
began to be applied. It was taken up by the Erste Brünner
Maschinenfabrik of Brünn (now Brno, Czechoslovakia) and
licences were taken out by several Continental and British firms
(see Fig. 43). Up to the period of the Great War several
hundreds of engines had been made for driving mills, fans and
electric generators, for winding from mines, for propelling

ships, and many locomotives including a goods locomotive on the North Eastern Railway were so fitted.

The advantages of the engine over the multi-cylinder horizontal engine are that it is more economical, occupies less floor space, is lower in upkeep because of the small number of working parts, and is as efficient as a triple-expansion engine. Economy is due to the reduction in initial cylinder condensation, resulting from the cylinder ends being kept at the boiler temperature, the reduction of clearance volume and surface and the high com-

Fig. 43. Uniflow steam engine.
Courtesy of Messrs Robey & Co. Ltd.

pression. Tests show that for engines developing between 500 and 1500 indicated h.p. with a boiler pressure of 175 lb. and 180° F. superheat, the steam consumption is the remarkably low one of 10 lb. per indicated h.p.

Seeing that expansion is carried out in one cylinder only, the ratio of expansion must be high, and this requires an early cut-off, say 8 to 10 per cent. This implies a high initial load on the cylinder head and piston, and consequently exceedingly heavy reciprocating parts have to be used, and these have to be carefully balanced. All this calls for a high standard in design and workmanship (see Fig. 44). A relieving gear to prevent the compression arising above a predetermined limit must be provided.[1]

---

[1] Cf. Stumpf, J., *Una-flow Steam engine*, 1912; and Perry, T. B., "The uniflow steam engine", *Proc. Inst. Mech. Eng.* 1920, p. 731.

The temperature gradient along the cylinder is so marked that to prevent trouble, due to expansion of the metal, cylinders have to be bored barrel-shaped when cold, i.e. slightly larger in diameter in the middle than at the ends. The uniflow engine is a gallant attempt on the part of the reciprocating engine-makers to hold the field against the advance of the turbo-alternator and the electrical drive, and it still maintains its ground for such situations as we have indicated above.

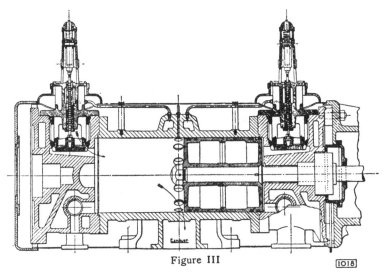

Figure III

Fig. 44. Section through cylinder of uniflow steam engine.
Courtesy of Messrs Robey & Co. Ltd.

Lubrication and packing. As pressures increased difficulties arose with lubrication and with the packing of pistons and rods. Vegetable lubricants like rape, otherwise colza oil, castor oil, etc., and animal lubricants like tallow, at pressures over, say, 80 lb. per sq. in. begin to be decomposed and acid is formed which attacks the metal of the cylinder and other parts with which the lubricant comes into contact. With further rise in temperature the lubricants carbonize and become useless. Fortunately relief was at hand, following upon the commercial exploitation of petroleum in the United States by "Col." E. L. Drake in 1858. Lubricating oils obtained from petroleum by distillation were found to be stable, and these mixed with small proportions of fatty acid are in use to-day.

As regards packing, hemp and tallow had to be given up for the same reason. Happily in the third quarter of the nineteenth century asbestos was exploited and became increasingly tractable, so that this material interwoven with brass wire into a fabric, impregnated with vulcanized indiarubber or graphite and made up into rope cord or block form is found capable of standing up to the highest temperatures.

For cylinders, besides modifications of Barton's metallic packing already described, there is the arrangement of two split rings forced out radially and axially by one or more sets of volute or leaf springs. The objection to rings is that they can only with difficulty be adjusted for wear. The simplest packing ring, very frequently found, is that patented (No. 767) by John Ramsbottom in 1852 for locomotive cylinders. Rings usually in pairs with the gaps breaking joint and a peg to prevent them sliding circumferentially are sprung into a groove in the piston. These can be seen in Fig. 44. Mention may be made of the "Clupet" ring patented in 1919 by B. Clews and H. M. Peterson, which is made by an ingenious milling process, and when sprung into position there is no gap; thus leakage is prevented and friction is reduced.

For piston rods, the earliest improvement in packing was the "lantern brass" of Jonathan Hornblower, 1781; he divided the packing, inserted in the intervening space two rings fitting both rod and stuffing box and held apart by three pillars, and into this annular space he introduced boiler steam. This "steam" packing has reappeared many times since. A modern metallic form of it is shown in Fig. 45, and it will be seen in position in Fig. 44. It embodies the combined improvements of several successive inventors since about 1875. It is duplex, i.e. there is an outer block packing in which the steam acts radially, and an inner cone packing in which the action is axial. This action takes place automatically, so that pressure is relieved on the exhaust stroke and consequently wear is slight. The packing permits of slight lateral and angular

displacement of the rod from any cause. It will best be understood from the legend given below the figure.

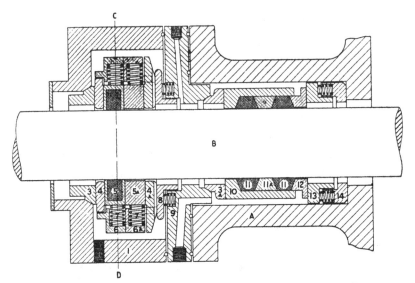

Fig. 45. Duplex metallic piston-rod packing, horizontal type.

Courtesy of the United States Metallic Packing Co. Ltd.

A. Stuffing box.   B. Piston rod.
1. Casing with steam inlets.
3 and 3 A. Ball-seated rings.
4 and 4 A. Ball-ring sliding plates.
5 and 5 A. Packing and guide blocks.
6 and 6 A. Horn rings enclosing seven block springs.
8. Spring cover plate.
9. Back plate with steam inlet.
10. Vibrating cup.
11 and 11 A. Cone rings.
12. Follower.
13. Spring cover.
14. Springs in spring holder.

For medium and low pressures the block packing alone is used;
for pressures below the atmosphere the duplex packing is used,
but the two parts are placed face to face and act in opposite
directions. The block part is open to the atmosphere which
actually tightens it to prevent the passage of air inwards.

We must now take up again the development of the boiler.

CHAPTER IX

LAND BOILERS, 1851 TO 1900

Improvements in the Lancashire type—Influence of marine practice—
Advances in the United States—Benson's boiler—Harrison's honeycomb
boiler—Water-tube, flash and coil boilers—Power stations—Superheaters—
Boiler efficiency.

THE second half of the nineteenth century opened, as
regards boilers, with the Lancashire type in the ascen-
dant. Improvements in details of construction, keeping
pace with the slow increase in steam pressure demanded, enabled
it to maintain its position and eventually take the lead; concurrently improvements in workmanship were made. It may be
said that in this half century the boiler-maker became a mechanical
engineer. Fig. 25 shows a longitudinal section and front views
of the boiler and its setting, to use the words of the inventor,
Fairbairn, "as I have been accustomed to construct it".[1] It
will be noticed that the boiler is set with a "wheel" draught;
since his time it has been more usual to set it with a "split"
draught, i.e. the hot gases from both furnaces unite and pass
beneath the boiler to the front, where they split and pass along
the sides to the chimney flue. The internal flues being long
cylinders were found to be liable to distortion and collapse.
Fairbairn carried out a series of elaborate experiments and
investigations on these cylinders and got over the difficulty
by the provision at intervals of strengthening rings as shown.
More effective were the ingenious devices patented by Daniel
Adamson and Leonard Cooper in 1851 (No. 14,259), by Thomas
Hill in 1860 (No. 258) and by James Noah Paxton in 1885
(No. 1275) seen in Fig. 46. Another expedient was to insert
in the flues tapered water tubes which not only acted as stays
but also added to the heating surface and promoted circulation; these tubes were patented by John and James Gallo-

[1] *Mills and Millwork*, 1861, Pt I, p. 256.

way of Manchester in 1851 (No. 13,532). Another way of strengthening the flues was by corrugating them—the invention of Samson Fox of Leeds in 1876. They were made at first by hammering but in 1878 Messrs Schultz and Knaudt of Düsseldorf, who had bought Fox's foreign rights, constructed a mill for rolling the flues. Such flues were shown at the Exhibition in that city in 1880. Since then, variants of the corrugated flue, such as D. B. Morison's "suspension" furnace patented in 1892 (No. 4806) and W. Deighton's flue patented in 1895 (No. 13,809), have been brought out; these have received widest application in marine practice. Longitudinal stay bolts, and gusset plates to connect the flat ends of the boiler to the shell, were ordinary practice with Daniel Adamson as early as 1856

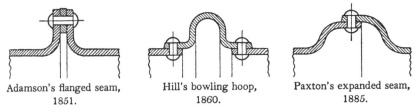

Adamson's flanged seam,   Hill's bowling hoop,   Paxton's expanded seam,
1851.   1860.   1885.

From their respective patent specifications.

Fig. 46. Lancashire boiler furnace seams.

for a pressure of 110 lb. per sq. in. Flanging the end plates instead of using angle irons to make the connection was introduced about 1866. The latest practice is to dish the end plates as well as flange them thus obviating the necessity either for stay bolts or gusset plates; the credit for this improvement is due to John Thompson of Wolverhampton in 1905.

As regards materials, the use of mild steel instead of wrought iron, because of its greater tensile strength, was permitted in 1865, after being regarded at first, and rightly so, as untrustworthy. As regards construction, drilling instead of punching holes in shell and butt straps, and machine riveting, came in during the next decade.

The vogue that the Lancashire boiler enjoyed is due to the fact that at no period was there a sudden demand for higher pressures, so that it was possible to satisfy increasingly exacting conditions as they arose by the improvements just indicated. The limit has now been passed, but there are rôles that the boiler can fill usefully for many years yet.

Several variants of the boiler have been introduced; one such is the "Yorkshire" boiler patented by William Herbert Casmey in 1906 (No. 22,925); in this the flues incline upward and increase in diameter from front to back, the diameter of the shell being proportionately greater and the length less than in the Lancashire type.

Alongside this steady progress, much thought and energy were being devoted to the improvement of the water-tube boiler. At first the departure was in the domain of marine practice and the reason was largely an economic one, for at sea the cost and weight of fuel to be carried are vital factors as they decide the radius of action of the vessel. We have to point to the water-tube boiler patented by J. M. Rowan and fitted in S.S. "Thetis" in 1858 as the pioneer in this direction.[1] Assisted by his son F. J. Rowan and his partner T. R. Horton, he continued to develop this boiler but with limited success.

For the first steps on land, we must turn to the United States, where the overflowing of the population into the Middle West, the change over from a predominantly agricultural economy to one combined with manufactures, the spread of trunk railways, and lastly the consolidation of the nation by the Civil War, 1862–5, stimulated engineering into great activity. In regard to boilers there was little or no respect for past experience. The tank boiler, so deeply entrenched in the old country, never took root in the States. There was a realization of the economy of high pressure, a willingness to experiment and take risks so that it is not surprising that a large number of sectional and water-tube boilers originated there. Another reason possibly was that the care and insurance of boilers were not tightened up in the States to the same extent as in Great Britain. Boiler explosions were not infrequent and loss of life was accepted with as much equanimity as is the toll of the motor car to-day. Hence the type of boiler that was least dangerous when it did

[1] Smith, *Short history of marine engineering*, 1938, pp. 197 *et seq.*

explode was bound to find favour. Of such were the sectional and the water-tube types.

A few of these boilers, to which close application and thought have continuously been devoted, still survive in an improved form. Others which came out with a flourish of trumpets, although they possessed meritorious features and were made in considerable numbers, were doomed by the verdict of experience, and they have disappeared so completely that their names even are unknown to the present generation.

Among the latter class was a boiler that embodied the then novel principle of forced circulation of the water—the invention of Martin Benson of Cincinnati, Ohio, and patented by him in this country in 1858 (No. 1903) more particularly for marine purposes. It consisted of small tubes about 1 in. internal diameter, built up zigzagged in series to form a vertical element. A number of these elements were placed side by side over the grate. The lower ends of the elements were connected to an inlet header from the circulating pump and the upper ends to an outlet header. The headers were connected to a vertical steam-and-water drum outside the boiler setting. A test boiler of 340 sq. ft. heating and 9 sq. ft. grate surface, working at 80 lb. pressure, was constructed by Messrs R. and W. Hawthorn, Leslie & Co. of Newcastle-upon-Tyne in 1859–61 for land use.[1] The pressure to be overcome by the circulating pump was 7–10 lb. per sq. in. and the best results were obtained when circulating 6 to 8 times the quantity of water that was evaporated; the water-steam mixture passed into the upper or steam space of the drum, and there the steam and water separated out. The boiler had some success in the United States. The principle of action was sound but the construction was defective, e.g. there does not seem to have been provision for cleaning. Nevertheless, one boiler is recorded to have worked for eleven years.

Another form of boiler that embodied meritorious features was the "honeycomb" sectional boiler of Joseph Harrison of

[1] *Proc. Inst. Mech. Eng.* 1861, p. 30, J. J. Russell on "Benson's High Pressure steam boiler", cf. 1859, p. 264.

Philadelphia, patented by him in 1859. It consisted (see Fig. 47) of an assemblage of communicating globular cast-iron chambers or cells, 8 in. external diameter, in sets of four. The

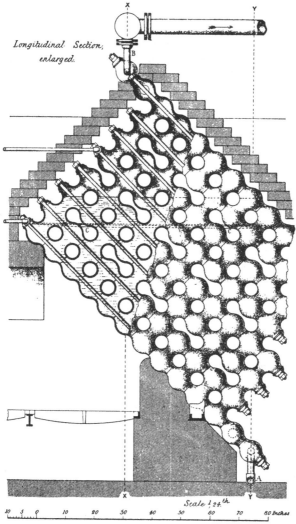

Fig. 47. Harrison's sectional boiler, 1862.
From *Proc. Inst. Mech. Eng.* 1864.

ends were machined and the sets were held together metal to metal by end covers and tie rods. In this way a boiler of any required amount of heating surface could be built up. The

assemblage of cells was placed at a slope of between 25° and 40°
to the horizontal in a brickwork setting, the grate being at the
lower end and the steam pipe at the top. Baffles ensured the
distribution of the hot gases among the cells. Our illustration[1]
shows a boiler constructed in this country. Trouble was experienced from leakage owing to differential expansion of the
wrought-iron tie rods and the cast-iron cells. The boiler had
considerable vogue but gave way before the pipe boiler.

Great Britain, never far behind, brought forth the next boiler
deserving of notice—that patented by Joseph Twibill in 1865
(No. 2431), and we mention it because he was the first, as far

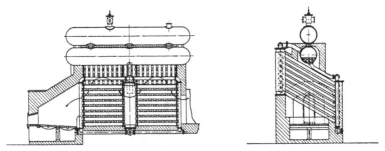

Fig. 48. Twibill's straight water-tube boiler.
From his patent specification, 1865.

as we have ascertained, to combine in a sectional boiler straight
tubes at a slight inclination to the horizontal (see Fig. 48), a
disposition that has proved to be, along with that of tubes at
a slight inclination to the vertical, very efficient. Incidentally
it may be remarked that every possible disposition and contour
of tube have been tried at one time or another but experience
has brought us back to the two dispositions mentioned. Twibill's
tubes were of wrought iron connected at the sides with stand
pipes having horizontal cross connections, joining to a horizontal
steam drum; this in turn was joined to a corresponding steam
drum above. There does not appear to have been any provision
for cleaning. Incidentally it may be said that it was in this
respect that most water-tube boilers came to grief. In those days

[1] *Proc. Inst. Mech. Eng.* 1864, p. 61.

it was thought almost "faddy" to filter boiler-feed water, while to soften it had hardly been contemplated. Nostrums were on sale to cause deposition of solid matter as mud in the economizer rather than as scale in the boiler; nevertheless, cases of boiler tubes having to be cut out because of being incrusted solid were common. Twibill was engaged in the manufacture of feed-water heaters or economizers and the influence of these on water-tube design deserves remark.

A water-tube boiler that created much stir in its day was that patented between 1861 and 1878 by Loftus Perkins (1834–91), grandson of the more famous Jacob. It consisted of layers of horizontal tubes crossing each other at right angles and joined to one another by screwed nipples. The boiler was tested to 2000 lb. per sq. in. and the working pressure was 500 lb. Distilled water was used, possibly the first instance of such use. The curious thing about the boiler was that there was no apparent course for the water to take to circulate, yet it worked. Pressures of over 200 lb. per sq. in. at that time were considered unsafe and only a few of these boilers were made.

The first sectional boiler made in this country commercially was that of John and Frederick Howard of Bedford, patented in 1868 onwards. If only because their pioneer work familiarized users with such boilers, they are deserving of remembrance. Their boiler (see Fig. 49) consisted of tiers of wrought-iron tubes 9 in. outside diameter at a slight inclination to the horizontal closed at the lower front ends and at the upper ends connected by nipples screwed into the flat side of D-shaped vertical tubes or headers. Each tier consisted of five tubes and these were usually set in brickwork in banks of five. Each header was joined to a steam receiver which in turn was joined to the steam main. Pressures up to 140 lb. per sq. in. could be sustained. Our illustration[1] is of a boiler modified for firing by blast-furnace gas at Lackenby Ironworks, Middlesbrough, in 1871. There were five of these units each of 50 h.p. working at 90 lb. pressure. A drawback was that circulation was defective and the tubes were found to be of unnecessarily large diameter.

[1] *Proc. Inst. Mech. Eng.* 1872, p. 274.

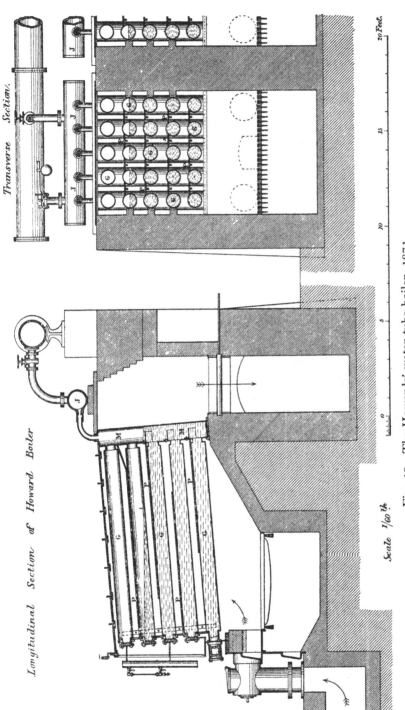

Transverse Section.

Longitudinal Section of Howard Boiler

Scale 1/60 th.

Fig. 49. The Howards' water-tube boiler, 1871.

From *Proc. Inst. Mech. Eng.* 1872.

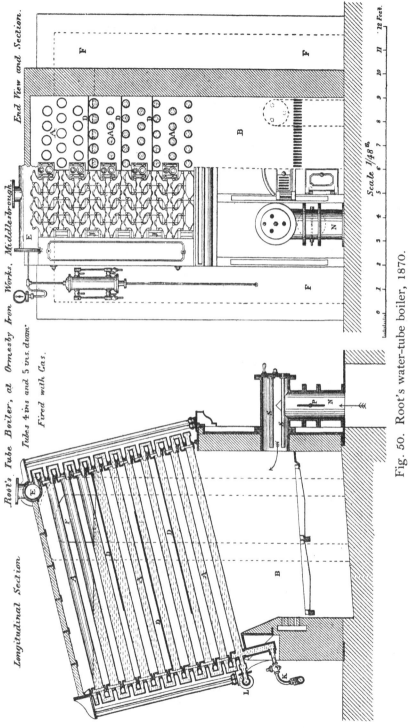

Root's Tube Boiler, at Ormesby Iron Works, Middlesbrough.
Tubes 4 ins and 5 ins diam.
Fired with Gas.

End View and Section.

Longitudinal Section

Scale 1/48

Fig. 50. Root's water-tube boiler, 1870.

From *Proc. Inst. Mech. Eng.* 1871.

Licences for making the boilers were granted and between 600 and 700 were made.

Reverting now to boilers that have by steady improvement kept to the front and are still represented to-day, we instance that one brought out in the United States in the 'sixties by John Benjamin Root and patented in Great Britain in 1867–9. It is of the small tube type, the tubes being inclined 1 in 3 to the horizontal in a bank connected at the upper end to a cross drum for steam. The circulation was internal to the bank itself by staggered connections. The boiler was taken up in England in 1870 by several firms. Our illustration[1] (see Fig. 50) represents the boiler as it was made in 1871, and arranged for firing by blast-furnace gas at Ormesby Ironworks, Middlesbrough. There were two boilers working at 70 lb. pressure. The boiler was likewise introduced into Germany, where the first works for making these boilers or indeed a water-tube boiler of any description was established in 1873 at Kalk near Cologne by Walther & Co. This firm was followed, as patents expired, by other makers, but with improvements of their own. The boiler has continued to be made largely on the Continent, although little semblance to the original, beyond the arrangement of inclined tubes, persists.

Another sectional boiler that has survived is that of George Herman Babcock (1832–93) and Stephen Wilcox (1830–93), patented by them in the United States in 1867. In reality this was based on an earlier design of Wilcox's patented in conjunction with O. M. Stillman in 1856 (see Fig. 51). The latter may best be described as consisting of an enlarged firebox crossed by sinuous water tubes inclined slightly to the horizontal, uniting front and back tube-plates or water-headers. This design was not proceeded with but foreshadowed the later design in which the tubes were straight to admit of being cleaned. These tubes were situated one above another with connecting headers each formed of a single casting, thus constituting a bank of tubes. A number of these banks side by side were connected to a drum above which served as a water-and-

[1] *Proc. Inst. Mech. Eng.* 1871, p. 229.

steam reservoir. Hand holes were provided opposite the ends of the inclined tubes for cleaning purposes. Internal tubes were at first placed in the inclined tubes with a view to promote

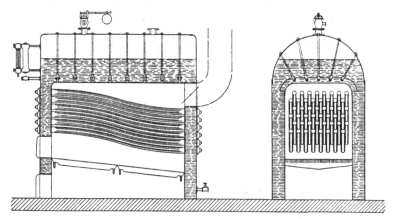

Fig. 51. Wilcox and Stillman's water-tube boiler, 1856.
From their U.S. patent specification.

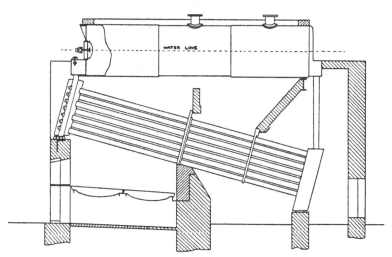

Fig. 52. Babcock & Wilcox's water-tube boiler, 1876.
Courtesy of Messrs Babcock & Wilcox Ltd.

circulation, but this device was quickly dropped as it interfered with cleaning. The design was sound (see Fig. 52); it remained standard for many years and was moreover susceptible of

improvement. This has gone on steadily and to-day, as we shall
see later, the boiler is still to the fore.

A distinctive type of vertical water-tube boiler is that known
by the name of its inventor, Edward Field, who with Moses
Merryweather and others patented it in 1862 (No. 2956). It is
composed of a number of elements (see Fig. 53), each of which
consists of two concentric tubes usually
placed vertically. The outer tube has a
closed end and is exposed to the fire so
that the steam rises in the annular space
between it and the inner tube while water
to replace that boiled off comes down the
inside tube. The boiler was designed for
steam fire-extinguishing engines where
rapid steaming was imperative.

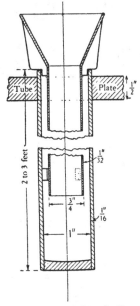

The element need not necessarily be
vertical. This has led to other applications
where the tubes are inclined and assembled
in double-chambered headers, one part of
the chamber serving to bring down the
water and the other to lead away the steam
to a drum or collector above. Practical
difficulties were met with in the joint
between the inner tube and the header;
eventually these difficulties were cleverly
overcome, between 1891 and 1910, by
the joint efforts of G. N. L. and P. E. J.

Fig. 53. Field's boiler
tube, 1862.

Drawing in the Science
Museum, London.

Niclausse of Paris. The boiler was used both for power stations
and on board ship, but it is difficult to empty and still more to
clean and these drawbacks have led to its virtual disappearance.

A boiler that should have been mentioned earlier is the
"flash" boiler, because it was one of the first to be patented.
Its principle is simple—water, which must be pure, e.g. distilled, is sprayed against red-hot metallic surfaces and is at once
turned into steam, in fact the feed is adjusted to give the exact
amount of steam required. John Payne as long ago as 1736
patented (No. 555) a flash boiler but it was not tried out till

1827 (Patent No. 5477) by Jacob Perkins, who pumped hot water through square hollow cast-iron bars kept at a red heat in a furnace and so obtained superheated steam at a pressure of 1500 lb. per sq. in.

There was a rebirth of the flash boiler in this country in 1894–5 when the Legislature allowed the mechanically propelled carriage to reappear on the road after an eclipse of sixty years. Some of these cars were steam propelled and of such the boilers were of the flash type, oil-fired. The best known was that of Léon Serpollet (1859–1907), who had developed it from 1887 onwards in France, where restrictions such as those in force in this country never existed. However, the boiler has not been used for stationary purposes and therefore does not concern us.

The coil boiler, like Rumsey's already mentioned, is one that has had a fascination for inventors. We may mention that of Julien Belleville of 1856 and, later, that introduced in 1893 by James Brown Herreshoff (1834–1930) and John Brown Herreshoff (1841–1915). In this boiler there was a single coil of very great length; in the "Climax" boiler patented in 1886 by T. T. Morrin there are a large number of coils expanded at both ends into a vertical shell or drum; the coils overlap and are staggered in a spiral formation. The whole is placed inside a brick chamber and the hot gases traverse the space between it and the drum. These boilers were made in large sizes and were installed in power stations.

With the introduction of electric lighting from 1885 onwards and the establishment of power for generating the current, a demand arose for boilers of high-pressure and rapid-steaming capacity, requirements which the water-tube type successfully met, so that there has been great development not only in size of unit but in the use of solid fuel either in pulverized state or with stokers, and in the use of liquid fuel with burners; there has also been advance in feed-water heating and in superheating.

Black smoke has always been the bugbear of boiler-firing; it is particularly offensive in populous areas and legislation has been directed against it. Much can be done by correct firing, but it has been realized more and more that complete combustion is the true remedy and this can only be secured by the provision of a zone of very high temperature prior to reaching the boiler itself. This is effected readily in the water-tube boiler and examples will be found in subsequent pages.

Superheaters. Steam in the condition in which it is ordinarily evaporated is "wet", i.e. it contains water vapour in suspension, and when this comes into contact with the cool surface of the cylinder it condenses, causing loss of economy. If the steam is heated to a temperature beyond that due to its pressure—a practice formerly known as "surcharging" and nowadays as "superheating"—it will be "dry" and such condensation will not take place. Until the middle of the nineteenth century opinion among engineers about the value of superheating was divided. Gustav Adolf Hirn (1815–90) was the first, by investigating the heat-balance sheet of a steam engine, to establish the advantages of superheating. As usual, where economy in fuel is sought, it is in marine practice that we must look for advance. In 1856 John Weatherell introduced his plan of mixing superheated with wet steam and effected an economy. The practical application was retarded till metallic packing which could not be burnt out was available.

About 1890[1] the practice was reintroduced and is now general. The apparatus takes several forms, but in nearly all it consists of a nest of tubes in parallel; the position selected for it is in the hottest zone. In the water-tube boiler the superheater is incorporated in the design (see Fig. 75).

Notwithstanding the variation in construction and proportions of boilers, from Newcomen's onwards, the difference in their efficiencies comparatively small. Theoretically 1 lb. of coal should evaporate, depending on its calorific value, 12 to 16 lb. of water from and at 212° F.; in practice the amount is 8 to 10 lb., so that the evaporative efficiency is anywhere between 50 and 85 per cent. Higher efficiencies have been established in boilers to be described in a subsequent chapter.

[1] *Proc. Inst. Mech. Eng.* 1890, p. 134.

CHAPTER X

PHILOSOPHY OF THE STEAM ENGINE

Measurement and volume—Heat and temperature—Energy and force—
Mechanical equivalent of heat—Carnot's cycle—Working substances—
Degradation of energy—Heat-energy stream of a condensing engine—
Efficiency of engines.

AT the outset we saw that inventors who attempted to obtain mechanical energy from natural sources were simply groping in the dark along avenues that seemed to hold out some hope of success. Having hit upon the idea of condensing steam under a piston and allowing the pressure of the atmosphere to do work, considerable progress was possible without it being necessary to know much about heat or the properties of steam. It is fortunate for mankind that important advances in practice can be based on mere empirical knowledge, for otherwise further advances could hardly be made. It is, however, this kind of knowledge, based on ignorance, that gives to the practical man such an opportunity, unhappily, to decry theory. When there are many workers in a new field, however, there are sure to be one or two persons here and there who look beyond externals to discover first principles, whence the real direction of further progress is to be sought. So it was in the case of the steam engine.

The main directions in which research has been undertaken have been physical and consequently in fields somewhat alien to those of the engineer. Nevertheless, the share that he has taken in the work has been notable. It was obvious in the first place that it was desirable to examine into the properties of steam; to do so led to enquiry into the nature of heat and its measurement. The interdependence between the amount of fuel burnt under the boiler and the quantity of water the engine could raise with it, i.e. the "duty", became the accepted criterion of performance. The power that a steam engine could exert, i.e. the horse-power or rate of doing work, was a practical consideration with every possible purchaser and such a method of rating engines grew up as early as 1785. Gradually the idea of a relationship between the amount of fuel and the horse power that could be obtained from it became a familiar one. Looking below the surface of this relationship, force and energy became disentangled from horse power, and the idea grew stronger and stronger that it was the heat in the fuel that was correlated with the force or energy that was liberated in the engine. This was consummated in the discovery of the mechanical equivalent of heat. The output of energy in an engine was thus seen to depend on the limits of temperature between which the engine works, and to increase the energy, the limit must be extended; hence the necessity of higher and higher pressures. Lastly, it was realized that there is energy that is available and some that is non-available, and that degradation of energy can take place. We must now enlarge somewhat on these themes.

Different notions as to what steam really is were held in the early days. Triewald, for instance, in 1734, believed that steam was "nothing but moist air, heated to a high degree, every particle of air being surrounded by an incomparably thin membrane or coat of water, very much like a bladder". It was realized, at any rate by Watt, that steam was a gas[1] obeying Boyle's and Mariotte's Law more or less perfectly.

One of the first questions that was asked about steam was, what is the volume that will be evaporated from a given volume of water at ordinary temperature and pressure? Sir Samuel Morland (1625–95) estimated this volume at two thousand times that of the water whence it was obtained. Henry Beighton, whom we have frequently mentioned already because of his interest in the engine, made some experiments to ascertain this volume, and he communicated his results to Desaguliers. The latter gives the figure 14,000, but we cannot believe that this was the figure given to him by Beighton. Shorn of one nought,

---

[1] Steam is invisible and what we see and call steam is really condensed vapour of water. This we realize if we look at the funnel of a locomotive blowing off; the condensed vapour appears only when at some inches above the top; in the space between it exists as steam.

it is a close approximation to the truth, and we may charitably put down as a printer's error this nought too many.

No advance in knowledge was made for nearly half a century until James Watt, as we have seen, had his attention directed to the atmospheric engine. He tackled the problems involved in a truly scientific spirit and showed remarkable insight. He determined the volume of steam to be 1800 times the bulk of the water at boiling point whence it was evaporated, a closer approximation to the actual figure—1642—than had previously been arrived at. With a Papin's digester he took the temperatures at which water boiled at different pressures above the atmosphere and found that "when the heats proceed in an arithmetical, the elasticities proceed in some geometrical ratio". He passed steam into a quantity of cold water until the latter got boiling hot, and found it gained above one-sixth in volume; in other words, the water converted into steam had raised about six times its own weight of water from room temperature to boiling point. Beighton had made a similar experiment, but his observation was not quantitative. The reason for this unexpectedly high result was explained to Watt by Professor Joseph Black as being due to latent heat, a discovery that the latter had recently made, namely, that bodies on changing their physical state absorb or give out heat at the moment of change, without any change in sensible temperature. We can imagine how delighted Black must have been at having his doctrine thus independently confirmed so quickly after its formulation! Watt sensed the application of this knowledge to the atmospheric engine, and the result was his invention of the separate condenser. We ought to mention that in 1781 Watt repeated his experiments on the pressure and temperature of steam, and that his assistant, John Southern, experimented at different dates, 1797 to 1814, on the same subject. Watt's conclusion was that latent heat was constant.

Henri Victor Regnault (1810–78) at the expense of the French Government, and the Franklin Institute, in 1843, for the American Government, extended the measurements of pressures and temperatures up to 24 atmospheres. In 1847 Regnault found that latent heat was not constant, i.e. that latent plus sensible heat increased by 0·305 of a Centigrade heat unit for each degree of increase of sensible heat. He also found that steam does not obey Boyle's Law, since it is not a perfect gas. Professor H. L. Callendar (1863–1930), between 1915 and 1924, redetermined all the data and collected the results, which he embodied in steam tables which till recently were accepted by all engineers.

We now come to the wider and more embracing question of the relationship of heat to work—in other words, the amount of work that can be obtained from a given amount of heat. This conception, we believe, arose gradually in men's minds from the practice of engine testing, originated by Smeaton and Watt, followed by others, in which the quantity of water that could be pumped to a defined height by a certain amount of coal was ascertained and used as a criterion of the performance of one engine as compared with another. This "duty", as it was called in Cornwall, was reckoned as the number of pounds of water raised one foot high per bushel of coal. There is nothing scientific about it, and like horse power, it is merely a convenient way of rating an engine. Nevertheless, it does cause a reflective mind to focus on the idea that there is a relationship between weight lifted, otherwise work done, and the amount of coal, otherwise heat, that is required to do it.

At the end of the eighteenth century the accepted view about heat was the caloric or material theory. Caloric was supposed to be a "subtle elastic fluid which permeated the pores of bodies and filled the interstices between the molecules of matter. It could not be created, neither could it be destroyed, nor yet had it weight." It was Benjamin Thompson, Count Rumford (1753–1814), who, as the result of experiments on the heat evolved by boring cannon, carried out at Munich in 1798, put forward the first suggestion that heat was a mode of motion of the molecules of a substance. His conclusion was: "Anything which any *insulated* body, or system of bodies, can continue to furnish *without limitation* cannot possibly be a material substance; and it appears to me to be extremely difficult, if not quite impossible, to form any distinct idea of anything capable of being excited and communicated in these experiments, except it be motion."

Slowly the idea gained ground, but it remained for Nicolas Léonard Sadi Carnot (1796–1832) to be the first to evolve anything like a theory of the performance of the steam engine. To use his own words, in his *Réflexions sur la puissance motrice du feu*, 1824:

> Malgré les travaux de tous genres entrepris sur les machines à feu, malgré l'état satisfaisant où elles sont aujourd'hui parvenues, leur théorie est fort peu avancée, et les essais amélioratifs tentés sur elles sont encore dirigés presque au hasard.

Carnot held the view that

> La chaleur n'est autre chose que la puissance motrice, ou plutôt que le mouvement qui a changé de forme. C'est un mouvement dans les particules des corps.

Carnot laid down the principle that the only way in which heat does work is by descending from a higher to a lower temperature; that efficiency is independent of the working substance and depends only on the initial and final temperatures; that it can never exceed $\dfrac{T_1 - T_2}{T_1}$, where T_1 is the initial absolute temperature of the working substance (in this case that of the boiler) and T_2 the final temperature (in this case that of the condenser). Carnot's work almost escaped attention even by scientific men; by engineers it was unnoticed—indeed, had they seen it they would never have believed that it had any relevance to practice. He was before his time.

The new doctrine of heat germinated slowly for more than a quarter of a century, and many minds contributed to it. One of these must be mentioned, Robert Mayer (1814–78), a young physician of Heilbronn, who, in 1842, succeeded in getting published his memoir "Bemerkungen über die Kraft der unbelebten Natur", but it attracted no attention although it enunciated the principle that the energy of the world is constant. The most outstanding thinker and experimentalist of the day was James Prescott Joule (1818–89) of Manchester, who, under the influence of John Dalton, began experimenting in 1840 on the relations between mechanical, electrical and chemical effects. In 1843 he communicated to the British Association, to "an unwilling audience of six", his paper on the "caloric effects of magneto-electricity", but it received scarcely any attention.

In the same year, Ludwig August Colding (1813–88), a Danish engineer, in an essay submitted to the Royal Society of Copenhagen, laid down "that force is imperishable and immortal, and therefore when and where forces seem to vanish in performing certain mechanical, chemical, or other work, the force then merely undergoes a transformation and reappears in a new form, but of the original amount as an active force".

It remained for Joule to bring forward the subject again at the British Association meeting at Oxford in 1847, but again it bid fair to be passed over in silence, "if", as Joule said, "a young man had not risen in the Section and by his intelligent observations created a lively interest in the new theory". That young man was Professor William Thomson (1824–1907), afterwards Lord Kelvin, who, it should be understood, had heard of Carnot's essay from Paul Émile Clapeyron (1799–1864) when they were working in 1845 in the laboratory of Regnault. Thomson's account of the reading of the paper is slightly different:

I...felt strongly impelled to rise and say it must be wrong.... But as I listened on and on I saw that though Carnot had vitally important truth not to be abandoned, Joule had certainly a great truth and a great discovery, and a most important measurement to bring forward. So instead of rising with my objection to the meeting, I waited till it was over and said my say to Joule himself at the end.

Joule's work culminated in his memoir "On the mechanical equivalent of heat", communicated to the Royal Society by Faraday, June 21, 1849. Joule determined that the quantity of heat required to raise 1 lb. of water 1° F. at its maximum density 4° F. was equivalent to the work done in raising 772 lb. 1 ft. high—a quantity now known as the British Thermal Unit (B.Th.U.); this value was accepted for half a century. Recent redeterminations by Henry A. Rowland of Baltimore and others have resulted in making only a slight correction to it; to-day the

value of 778 ft. lb. is accepted. Joule's work confirmed the law
of the conservation of energy and established the dynamical
theory of heat. Kelvin became a lifelong friend of Joule and an
ardent disciple, with Hermann Ludwig Ferdinand von Helmholtz
(1821–94), Rudolf Clausius (1822–88) and William John Mac-
quorn Rankine (1820–72). They had an uphill fight before the
science of thermodynamics may be said to have been founded.
Even in the *Encyclopedia Britannica*, 8th edition, 1853, the article
"Heat" was still treated from the old caloric point of view, with
scarcely a hint afforded of there being any application to the heat
engine. It was Professor Rankine, in particular in his *Steam
Engines and other Prime Movers*, 1859, who first brought home
to engineers what the quantitative application of the new teaching
as concerns the steam engine meant to them. Deduced from the
principle of the conservation of energy of Joule, the first law of
thermodynamics is arrived at. Expressed in the particular terms
understood by the engineer, and not in the more general terms
used by the physicist, we say that: *When mechanical energy is
produced from heat, a certain quantity of heat goes out of existence
for every unit of work done; conversely when heat is produced by the
expenditure of mechanical energy, the same quantity of heat comes
into existence for every unit of work spent.* The second law of
thermodynamics, deduced from Carnot's reasoning, but not
formulated by Clausius till 1850, is usually expressed thus:
*It is impossible for a self-acting machine, unaided by any external
agency, to convey heat from one body to another at a higher temperature.* All motive power involves the existence of a hot and
a cold body, one to act as a boiler, the other as a condenser—
a source and a sink of heat.

Carnot's theorem, expressed almost in his own words, is as
follows: *When a working substance that has passed through a series
of conditions modifying its temperature is brought back to its
original physical state as to density, temperature and molecular condition, it contains the same quantity of heat that it contained
originally.* His cycle is, it will thus be seen, a reversible one—
ideal in fact—but when we come to mundane apparatus like the
condensing steam engine we find the heat passed to the condenser

distributed over a large quantity of cooling water and therefore it cannot be collected and restored to the cycle, in fact the heat is of very low grade and is for practical purposes lost. Rankine realized this and hence approximated to the ideal by employing the cycle which goes by his name and which differs only from that of Carnot by assuming that no reversible process is followed in the return to the boiler of the heat in the condenser water. Thus it was that in the third quarter of the nineteenth century, it had been established that the reciprocating steam engine was a particular form of the heat engine, and that the object of the engine was to get work done with the least expenditure of fuel. It was realized that in such an engine there is a working substance that serves as a vehicle by which heat passes through the engine, that the substances that can act as vehicles may be liquid or gaseous, and that the way in which they act is normally by change of volume. It was manifest that heat is let down from a high level of temperature to a lower one, and that while this is taking place heat disappears and does mechanical work. This can be visualized on the analogy of the water wheel, where water is being let down from a high level to a lower one and does work in the process. Hence it was seen that to secure efficiency the essentials to be aimed at were: the highest possible boiler pressure with consequent high initial cylinder pressure, expansive working, relatively short stroke, high speed of revolution to diminish condensation of steam on admission and low condenser temperature and pressure. A new race of engineers now grew up who applied the new knowledge, as we have seen, to the construction of new engines. Prominent among these was Peter William Willans (1851–92).

We can now enquire what is the order of the efficiency that we can attain in the reciprocating engine. We are helped greatly in this by a study of the diagram (Fig. 54), based upon Professor W. E. Dalby's work,[1] of the heat-energy stream in a condensing engine. Starting at the grate of the boiler with a stream 100 units wide, it will be seen (neglecting small amounts, which are, however, shown on the diagram) that 5·5 per cent is lost by radiation

[1] *Steam Power*, 1915, p. 44.

from the boiler and 11 per cent is carried off up the chimney; radiation from the feed-water heater accounts for 2·7 per cent, but hot-well and cylinder-jacket circulating water bring back 6·4 per cent, so that 85 per cent enters the cylinder. Here 14·9 per cent of the original heat disappears, being transformed into work. There is a loss of 2·45 per cent in radiation from the engine. After small losses and gains, 60 per cent of the original heat is passed away in the condensing water. The efficiency of the whole plant is 14·9, the efficiency of the engine alone is 18·9 per

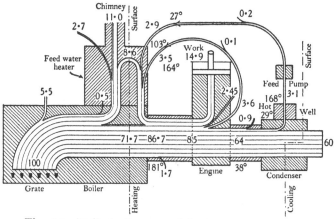

Fig. 54. Dalby's diagram of the heat-energy stream in a condensing engine.

Redrawn from his *Steam Power*, 1915.

cent, of the boiler alone 71·7 per cent and of the boiler plus the feed-water heater 80·3 per cent. This must be admitted to be a thoroughly bad showing, and the worst of it is there is so little scope for improvement. With the utmost efforts, an increase in efficiency of the order of 1 or 2 per cent is all that can be hoped for.

The position of the steam engine has not been materially altered by the advent of the steam turbine, to which we now turn. Its advent has enabled considerable advances to be made, principally by the possibility the turbine affords of raising the upper limit of temperature by the use of higher pressure and by superheat. There is none of the heating and cooling of the metal as in the reciprocating engine—each part of the turbine in working attains a temperature which remains constant. The turbine permits a method of regenerative feed-heating by "bleeding" steam at successive stages of the expansion to restore heat to the condensed water on its way back to the boiler; finally, the steam turbine permits expansion down to the lowest vacuum that condensing water can produce, expansion that is impracticable in the reciprocating engine owing to the huge volume that the steam then assumes.

CARL GUSTAF PATRIK
DE LAVAL
1845–1913

AUGUSTE CAMILLE
EDMOND RATEAU
1863–1930

CHARLES ALGERNON
PARSONS
1854–1931

CHARLES GORDON
CURTIS
1860

BIRGER
LJUNGSTRÖM
1872

PLATE VI. PIONEERS OF THE STEAM TURBINE

PART II

THE STEAM TURBINE

CHAPTER XI

KINETIC ENERGY OF STEAM AND PIONEERS OF ITS USE

Analogy of steam turbine with water turbine—Heron of Alexandria—Von Kempelen's mill—Trevithick's whirling engine—Branca's wheel—Parsons's upbringing—Partner in Clarke, Chapman & Co.—Patents axial flow steam turbine and dynamo—Partnership dissolved—Loss of patents—Radial flow turbine—Parsons recovers patents—Extension of turbine patent—Turbine blading.

HITHERTO we have been considering engines for making use of steam by its pressure alone, but there is another way in which steam can be employed. Taking water as the analogy we have the water wheel where we employ the pressure of the water due to its weight, and we have the water turbine where the kinetic energy of the water is used. So with steam: besides its pressure, which has been considered in previous chapters, we can use its kinetic energy; just as in the case of water we have two fundamental types of turbines, viz. the reaction and the impulse types, so with steam we have the same two forms of application. Curiously enough, when we come to enquire into the matter, we find that the application of the kinetic energy of steam was thought of much earlier than the application of its pressure, but as in so many cases, the first shall be last and the last first.

Early in the Christian era—the date is only approximately known—a writer named Heron, living in Alexandria, then the centre of commerce, the home of learned men, and the resort of students from all parts of the world, set out to record and describe in an encyclopedic manner the machines and apparatus existing in his day, with additions of his own invention. In one of his MSS. which has come down to us, entitled *Pneumatica*, Heron describes applications of heat to produce different motions such as the expansion of air to open a temple door, or the action of fire on liquid to pour out a libation on an altar. Heron describes also the use of a jet of steam to blow a horn or make a bird sing. Apparently the entire purpose of the employment of these means was, as we have said already, to create astonishment among the vulgar and to play upon their ignorance; that they might serve some useful purpose does not appear to have been thought of.

The reaction turbine. One of the toys described by Heron is the reaction wheel, or aeolipyle—the door of Aeolus—as it was subsequently termed. It consists of a hollow ball or globe on trunnions through one of which the steam is introduced from

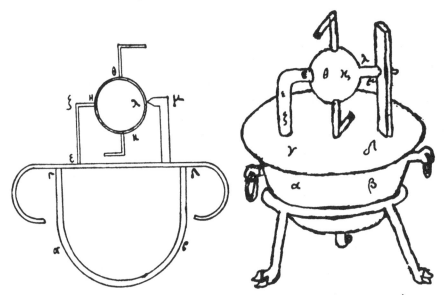

Fig. 55. Heron's steam engine,
c. A.D. 100.

From British Museum, Harleian MS. 5899,
fifteenth century.

Fig. 56. Heron's steam engine,
c. A.D. 100.

From British Museum, Burney MS. 81,
sixteenth century.

a boiler below; the ball is furnished with bent pipes or jets through which the steam is discharged, and thus the ball is caused to revolve. The MSS. of Heron are only known to us by copies made in the fifteenth and sixteenth centuries; these MSS. are illustrated by sketches which, however, differ among themselves in the different MSS. Such sketches may be degraded

copies of originals, or they may be interpretations evolved from
the inner consciousness of editors and scribes in the intervening
centuries. The sketches shown above are taken from MSS. in
the British Museum; there are, however, many other MSS. in
existence—and they help to bear out both of the suggestions
made above. However that may be, the mode of action is not
in any doubt and it should be noted
that it is the unbalanced pressure on
A, shown in Fig. 57, that causes the
rotation of the ball in the direction
shown by the arrow; actually the
apparatus would function better in a
vacuum than in the atmosphere, for
then the resistance of the air behind
A would be absent.

Fig. 57. Diagram to show
action of aeolipyle.

During the centuries that have elapsed since Heron's time
many attempts have been made to make a working machine out
of this toy, some of which have been successful. One such
attempt we must notice, because it came to the ears of James
Watt and at first alarmed him because he feared it might
jeopardize his reciprocating engine; that attempt was the invention that was patented April 11, 1784, by Wolfgang, Baron
von Kempelen of Presburg, Hungary, best known as the maker
of an automaton chess player. Dr Joseph Priestley's attention
was drawn to Kempelen's apparatus and the inventor's claims
were such that, could they have been substantiated, Watt's
engine might have been driven out of the field. Writing to
Boulton (May 11, 1784), Watt says: "Yesterday I heard from
Dr Priestley that Kempelen's engine is a Barker's mill[1] turning
in the air by the force of steam, which is very different from my
former idea of it and perhaps a worse thing for him. Before we
say much about its merits it will be best that he should specify"
(i.e. that he should patent it; this had, however, already been
done) "lest by talking about it we put him on improvements;

[1] Barker's mill is the simplest form of reaction water turbine, invented by
Dr Robert Barker, F.R.S., in 1743, cf. Desaguliers, *Experimental Philosophy*,
1744, II, p. 459.

for it is capable of them." Watt then proceeds to examine the invention critically and exhaustively thus:

I apprehend that the power is to be calculated in the following manner. Suppose the steam = to 30 inches mercury & the sum of the orifices = 1 sqr inch, then the propelling force will be 14 lb p^r inch when the machine is at rest, & as the velocity of stm issuing under said pressure is = 1800 feet p$^{r''}$ [i.e. second] & this machine is subject to the laws of undershot mills, the proper velocity for it is $\frac{1}{3}$ of y^e above i.e. 600 feet p$^{r''}$ & y^e effect will also be about $\frac{2}{3}$ of the power or 10 lb × 600 = 6000 lb 1 foot high p^r second. The quantity of steam employed will be 1800 ÷ 144 = 12·5 cubic feet steam. Now in one of our Engines 12$\frac{1}{2}$ cubic feet steam would raise 250 feet ∇ [i.e. water] to 10 feet high = to 15562 lb 1 foot high. This then would be the effect of the machine working in vacuo, compared with our engine, but by another theory it comes out that the effect might in that case be greater, but this second theory is so complicated that I cannot say I understand it thoroughly, nor do I think there is occasion because 600 feet p^r second seems to me an impossible velocity on account of the necessary friction of the machine. I should suppose 100 feet p^r second to be a maximum in that point, then the effect would be = 14 lb × 100 feet = 1400 lb 1 foot high p^r second by the same expense of steam viz 12·5 cubic feet.

But as he uses no vacuum & cannot without interfering with us, If he makes Steam = to 30 inches mercury, its density will be 900 & the velocity p$^{r''}$ will be 1290 feet. The proper velocity 430 feet p^r second, y^e effect 4300, & the quantity of steam 18 cubic feet because it is of double atmosphere density: He will find great difficulty in making boilers of any tolerable size which will be steam tight. This latter engine would be = to about 7 horses & would require a boiler $\frac{2}{3}$ as large as that at Chacewater, and in the case of 100 feet p^r second velocity working in vacuo.

The machine would be equall to 2$\frac{1}{2}$ horses, & would require a boiler = to y^e evaporating 26 cubic feet water p^r hour = to that of a 38 inch cylinder. Such a boiler would cost almost as much money as one of the engines which was of that power, but you will remember that if a velocity of 600 feet p$^{r''}$ be possible that the Kempelen [machine] working in vacuo would be = to 11 horses & the 38 inch cyl^r is = to 27 horses. So that you see the whole success of the machine depends on the possibility of prodigious velocities. The collar through which the steam is transmitted will have much friction, but that might be obviated in a great degree by making it turn in a collar of mercury which however if he has not thought of you will take care not to mention. As he has a very complete model of it ready made you may try on it what quantity of coals are burnt, ∇ [i.e. water] evaporated and work done, also if you can the density of the steam, but don't propose a steam guage if he has none. The fairest way of trying the power will be by pumping water or by raising a weight. In short without god makes it possible for things to move 1000 feet p$^{r''}$ it can not do much harm.

Watt reasons by analogy with a water jet, i.e. in a Barker's mill, and first calculates the velocity of a steam jet on the assumption that the jet is at one atmosphere pressure and discharges into a vacuum (of course there were no means at that time of getting such results experimentally). Kempelen was not doing this, however, and could not do so without infringing Watt's condenser patent. Actually Kempelen was discharging steam at one atmosphere pressure into the atmosphere. Now Watt had determined that the volume of steam at atmospheric pressure was 1800 times that of the water whence it came (see p. 175), and by Boyle's law at one atmosphere pressure the volume was 900 times (that is what he means by saying "its density will be 900"); consequently his reasoning was something like this: one atmosphere = 30 in. of mercury or 34 ft. of water; taking the formula $V^2 = 2gH$, H must be, in the case of steam, at a pressure of one atmosphere = 34 × 900, so that V, the velocity of the steam, is 1400 ft. per sec. (Watt's figure is 1290 ft. per sec.).

Based on this, Watt concludes that such a high speed of revolution of the machine is necessary as to be impracticable in the then state of the mechanic arts, and consequently that the machine can safely be neglected. The analysis of the problem is masterly and the conclusion irrefutable. Typical of the man is that his fertile brain throws out a suggestion for improving the machine, but desires that it should not be disclosed to the inventor for fear of adding to the value of the invention. Needless to say, neither Watt nor the world at large heard anything further about the machine.

Yet another attempt to make a working engine out of the

reaction wheel we feel we must record, because it was an emanation from the brain of that extraordinarily prolific inventive genius Richard Trevithick, apparently without knowledge that anything previously had been attempted in that direction. This engine he patented in 1815 and, as usual with him, he tried out the idea at once on a working scale—no wasting of time on models for him. His sketch shows a boiler supplying steam at 100 lb. pressure to a hollow axle on the end of which overhung double arms 15 ft. from nozzle to nozzle; the opening was $\frac{1}{2}$ in. by $\frac{1}{4}$ in. and the speed was 300 revs. per min. (see Fig. 58). The mere look of this gadget hurtling round at this speed must have been impressive, not to say alarming, and a source of danger. It is not surprising that it earned from the workmen the soubriquet of the "windmill", and the general belief was "that it was intended to throw balls at the French", with whom we were then at war. Trevithick realized that the defect of his "whirling engine", as he called it, just as Watt had done with that of Kempelen, was that he could not get a high enough velocity. Trevithick says: "I found that the great loss of power was ocasioned by not being able to drive the spouting arm at the extream end not above 200 feet per second, which bore a small propotain to the volosety of the spouting steam." At the pressure stated—100 lb. per sq. in.—the velocity of steam issuing into the atmosphere is about 2700 ft. per sec., and to obtain efficiency the arms should travel at slightly less than half this rate, so that the speed Trevithick actually attained was far too low. Once more, the idea being ahead of the mechanical arts of the day, it had to be dropped.

Nevertheless, the steam Barker's mill has been actually made into a workable machine and many sporadic examples—one might call them literally *tours de force*—have been made and used. Most of them have been as elementary as Trevithick's whirling engine, but other inventors have experimented with an engine comprising a number of Heron wheels working in series so as to distribute the pressure drop and obtain a more manageable velocity. The great drawback to the Heron type arises from the friction opposed to the revolving arms by the

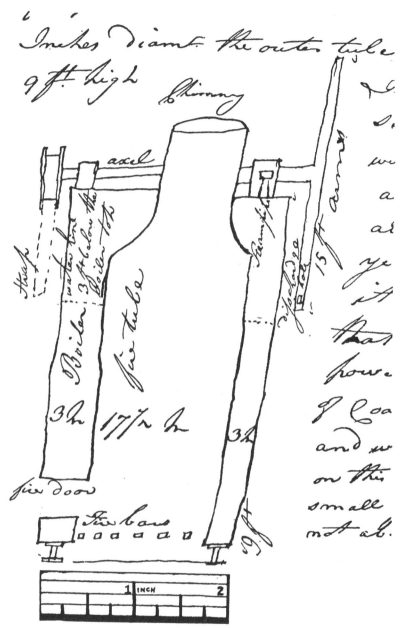

Fig. 58. Trevithick's whirling engine, 1815.

From the Enys Papers.

gaseous medium in which they work and by the centrifugal forces involved. Apart from the stress caused in the material, the limit to which is soon reached, the pressure of the steam at the point of escape is increased and is substantially higher than at the point of supply; this alone affects the efficiency. The steam consumption can be brought down to that of a reciprocating engine of like power but it is still high, so that there is no incentive to follow up the idea. We indulge the pious hope, however, that some day an inventor may arise who will make a success of what, after all, is the oldest steam engine in the world.

The impulse turbine. We mentioned above that there is a second way in which the kinetic energy of steam can be utilized, and that is by impulse, i.e. a jet of steam can be made to strike vanes on a wheel and cause the latter to revolve. Among others this idea occurred to Giovanni Branca of Loretto in Italy, and he described it along with a number of ingenious mechanical contrivances in his work published in 1629.[1] He shows a boiler in the form of a man's torso and from his mouth issues a jet of steam which impinges on the vanes of a horizontal wheel. Among other applications, the wheel is shown driving by gearing two pestles for pounding drugs in mortars (see Fig. 59), but the description reveals no apprehension of the speed at which the horizontal wheel ought to revolve. If ever made, it could only have been a toy. Needless to say, we hear no more of it.

We must pass over without description many further meritorious attempts to construct a steam turbine, such as that of James Pilbrow, 1843 (No. 9658) and that of Robert Wilson, 1848 (No. 12,026), and confine our attention to those that resulted in success. First and foremost we must mention the labours of the Hon. Charles Algernon Parsons (1854–1931). His work was the evolution of a practical reaction steam turbine. Sixth and youngest son of the Earl of Rosse, Past President of the Royal Society, young Parsons had enjoyed educational and

[1] *Le machine...del Signor Giovanni Branca, Cittadino Romano, Ingegniero, Architetto della Sta. Casa di Loretto,* MDCXXIX.

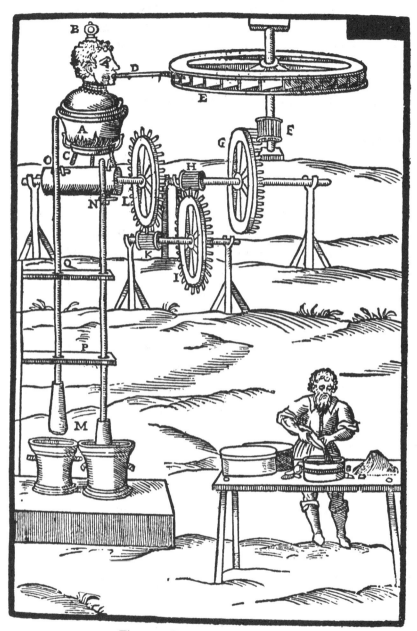

Fig. 59. Branca's steam turbine.

From his *Le machine*, 1629.

other advantages such as fall to the lot of very few. His early life has been sympathetically dealt with already in a biography[1] recently published, and we can therefore come at once to the year 1880 when, while still a pupil at Armstrongs' Works at Elswick, Newcastle-upon-Tyne, and familiar with the idea of obtaining power by the pressure of a gas upon a piston, he was led to interest himself, like so many others before him, in means for getting power from the kinetic energy of a gas. One of his ideas was to apply a jet of air for the propulsion of a vessel, and at the Heaton Works there is preserved a model of a boat with the propeller under the keel; the propeller is surrounded by a double shrouding, enclosing blades on which a jet of air was made to impinge. Following upon this and other experimental work, he conceived the steam turbine by realizing the analogy between the flow of steam under small differences of pressure and the flow of an incompressible fluid like water in the hydraulic turbine. Fortunately, we have Parsons's own statement as to how this conception came about, made many years after it is true, in 1922, in the course of his Presidential Address on "Motive Power" to the Birmingham and Midland Institute. These are his words:

In commencing to work on the steam turbine in 1884, it became clear to me that in view of the fact that the laws for the flow of steam through orifices, under small differences of head, were known to correspond closely with those for the flow of water, and that the efficiency of water turbines was known to be from seventy per cent. to eighty per cent., the safest course to follow was to adopt the water turbine as the basis of design for the steam turbine. In other words, it seemed to me to be reasonable to suppose that if the total drop of pressure in a steam turbine were to be divided up into a large number of small stages, and an elemental turbine like a water turbine were placed at each stage (which, as far as it was concerned, would be virtually working in an incompressible fluid) then each individual turbine of the series ought to give an efficiency similar to that of the water turbine, and that a high efficiency for the whole aggregate turbine would result; further, that only a moderate speed of revolution would be necessary to reach the maximum efficiency.

[1] Rollo Appleyard, *Charles Parsons, His Life and Work*, 1933.

Such was the brilliant conception. To carry it into effect a series of turbines[1] consisting of a ring of blades were fixed on a rotor within a fixed cylinder whence projected inwards rows of blades alternating with those of the rotor. Steam flowing parallel to the axis of the machine along the annulus between the rotor and the cylinder supplies the first turbine, and the exhaust from that is the supply of the second and so on, the pressure thus imparted being expended in driving the rotor blading. The pressure drop between each turbine being small, the velocity is within practical limits of shaft speed. To allow for the increase in volume of the steam as its pressure falls, the pitch and length of the blades are increased towards the exhaust end. To counteract end thrust, one right and one left set of turbines were provided and the boiler steam introduced centrally between them. Nor was this all; there were the mechanical problems involved by the high speed, amounting to thousands of revolutions as compared with the hundreds in the case of reciprocating engines. Shall we say without irreverence that the time had at long last arrived when, to use the words of Watt: "God made it possible for things to move a thousand feet per second"? These problems, the centrifugal forces, the bearings and their lubrication, the governing, were thought out and provided for. This was the construction shown in Parsons's world-famous patent of 1884, April 23, No. 6735, for "Improvements in rotary motors actuated by elastic fluid pressure and applicable also as pumps", a patent comparable in importance to that of Watt of 1769 for the separate condenser.

The idea of the turbine from the beginning was that of a machine that should drive directly the newly-introduced dynamo or electric generator, which, as we have seen, demanded a speed of about 1200 r.p.m. This while enormous for the reciprocating engine was small for the turbine, the speed of which was more than ten times as great, so that Parsons, alive to this fact, had to design a new generator. For this he took out a patent at the

---

[1] The nomenclature of the steam turbine adopted in these pages is as follows: Axial, not parallel, flow; Radial, inward or outward, flow; Cylinder, not casing nor stator; Rotor, not drum, shaft nor spindle.

same time as that for the turbine, viz. in 1884, April 23, No. 6734, for "Improvements in electric generators and in working them by fluid pressure". We cannot pursue this aspect of his work further than to recall what Dr Gerald G. Stoney, his life-long colleague and friend, says about these beginnings:[1] "Few better examples of Parsons's intuition can be given than the fact that the proportions of his first turbo-dynamo designed 53 years ago when electrical knowledge was so rudimentary, could, so far as I know, be hardly improved upon today and certainly could not have been for more than a quarter of a century after the machine was constructed."

Incidentally we ought to mention that Parsons had left Armstrongs in 1881, and for two years was with Messrs Kitson at Leeds, making himself a nuisance apparently owing to his overflowing inventive genius. In 1883 he joined the firm of Clarke, Chapman & Co., of Gateshead-on-Tyne, as junior partner in charge of the electrical engineering department, and under their agreement any patents taken out by the partners were to be the joint property of the firm and the partner concerned.

Parsons lost no time between the applications for his patents and the building of the first turbine and dynamo. This historic machine is illustrated in Pl. VII. The construction is deserving of close study, for it was the starting-point whence the vast development we know to-day has sprung. The machine was of the double-flow type, steam entering the cylinder at mid-length and exhausting at either end, thus eliminating end thrust, but the exhausts are brought together to a corresponding mid-length outlet. The blading consisted of slots cut at an angle of about 45° on the edge of gun-metal rings threaded on the rotor shaft. Alternating with these rings and fixed inside the cylinder, which was split axially, were half rings of similar blades. The form of the blading was dictated by the workshop facilities available at the time, and this was the case down to 1896. It is wonderful that with such crude shapes the results obtained were as good

[1] In his Parsons Memorial Lecture, November 25, 1937, *Journ. Inst. Elec. Eng.* LXXXII, p. 248.

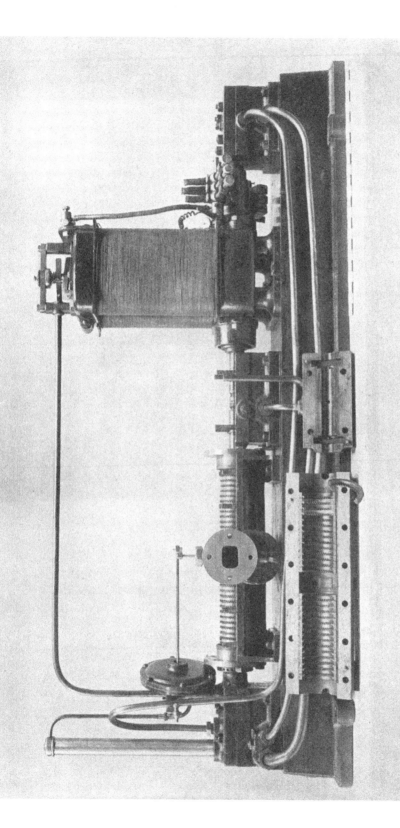

PLATE VII. PARSONS'S ORIGINAL STEAM TURBINE AND BIPOLAR DYNAMO 1884

Preserved in the Science Museum, London

as they were. There are 28 rows of blading, 14 fixed and 14 moving; in other words, the expansion was completed in 14 pressure steps. The rotor shaft journal ran in a long bush embraced by a large number of washers, forced together by a helical spring; one washer fitted the casing and the next the bush, and so on alternately. Thus the shaft could find its own alignment should its centre of gravity not be at its geometrical centre, which was almost inevitably the case; vibrationless running was thus ensured. The bearings were continually lubricated with oil from the pump. The governing was by throttle, this being actuated by an ingenious device, partly mechanical and partly electrical. The machine developed about 7·5 kW. at 100 volts and at a speed of 18,000 r.p.m. The blade velocity was 250 ft. per sec., with saturated steam at 80 lb. per sq. in. at the boiler, and exhausted at atmospheric pressure; the steam consumption was naturally high, being about 130 lb. per kWh.

The dynamo followed the construction of the first patent specification; it is a bipolar shunt-wound machine capable of an output of 75 amps. at 100 volts at the speed stated. After sixteen years of useful service this first turbine and dynamo have deservedly found an honoured home in the Science Museum, London.

This pioneer machine was at once successful and proved to be an exception to the rule that an invention but rarely springs complete from a single brain, like Minerva from the head of Jupiter. This is to be explained by Parsons's remarkable capabilities, about which Dr Stoney has this to say:

Parsons had an extraordinary intuition on all matters connected with design. No matter how difficult or novel the problem, he seemed to know instinctively how to solve it. Furthermore, he was often able to proceed directly to the best solution. . . .

Not only was he endowed with a thoroughly practical type of mind, but he had also great mechanical aptitude. He was perfectly familiar with all workshop processes and had unusual skill with his hands. He took a delight in making models, not as ends in themselves, but as means of studying some problem in which he was interested. These models were rarely very elaborate, but so well were they adapted to the purpose in view, and so great was Parsons's ability as an experimenter, that he was able to obtain very accurate results from quite crude apparatus.

The close parallel that exists between Parsons and Watt at once occurs to the mind. As to Parsons's mental powers, Dr Stoney observes that Parsons, although he possessed high mathematical ability, as is proved by the fact that he was eleventh wrangler in the Mathematical Tripos at Cambridge, "was rarely, if ever, known to make use of formal mathematical reasoning for the solution of any problem". Possessed as he was of a scientific mind and educated in scientific habits of thought, his results were the outcome of these rather than of deliberate calculation. He was never known to use a slide rule, and it is doubtful if he possessed one; "he seemed to have no need for such a thing", nor did he make use of graphic aids, but worked out results in his head; how he did this he never explained, even if he could have done so. "It never seemed to him that his powers were exceptional."

Perhaps we should say that Parsons's mind was of the synthetic order and that he could extract the kernel from mathematics without effort. On this point Dr Stoney has this further to say:

By analogy with the flow of water Parsons realized that, for small differences of pressure, the velocity of steam escaping from a nozzle could be represented by an expression of the form $V^2 = 2gh$. For water of course h is the static head in feet. In steam if H is the "homogeneous head" $(H = 144pV)$, it follows that $h = H\dfrac{dp}{p}$ and $V^2 = 2g H \dfrac{dp}{p}$. The great advantage of dealing with H in place of either pressure or volume is its relative constancy along a turbine. In a modern turbine the pressure may vary in the ratio of about 1000 to 1, whereas the corresponding variation of H will only be about 2·5 to 1. In fact, in the early non-condensing turbines, H only varied by about 15 per cent. Parsons certainly made much use of the function H, of which he was very fond.

Another example of this synthesis of mind is the birth, at a much later date, of the well-known "Parsons coefficient". Dr Stoney says:

I well remember the day when he sprang upon us what is now universally known as the "Parsons coefficient" K, with no more explanation than that it had the value $K = NR^2D^2 \times 10^{-9}$ where N is the number of rows [of blades] on the cylinder or spindle, R is the number of revolutions per minute and D the mean diameter of the blading in inches, a summation being made for the various stages of the turbine. At that time all we knew was that K was a figure related both to the turbine efficiency and to the ratio of blade speed to steam speed, and it took a good deal of cogitation before we discovered that this ratio u/c was equal to $\sqrt{\dfrac{K}{1\cdot3\,\Delta H}}$ for a reaction turbine, where ΔH denotes the heat drop. I have often wondered what connection there was in Parsons's mind between efficiency, velocity ratio and heat drop. He never told us, but accepted the above formula when it was suggested to him.

But to resume the thread of the story of the practical development of the steam turbine. The pioneer machine was quickly followed by others. One unit was shown at the Inventions Exhibition of 1885, but hardly a single engineer at that time looked upon it as more than a clever toy, much less as a serious competitor of the reciprocating engine; for one thing, its steam consumption was so high. However, there were plenty of openings for it as far as it had got, e.g. it came into extensive use for isolated lighting plants on shipboard, etc. In the year 1888, the first installation in a public power station, that of the Newcastle and District Electric Lighting Company, took place. This was a 75 kW. turbo-alternator, and by 1889 four such sets had been supplied; these were the first alternators to be put into use, all those previously supplied being direct current. Up to this date some 300 steam turbines had been supplied, the largest of which were the ones just mentioned.

Now occurred what we may justly term a turning point in turbine history. Differences of opinion arose between Parsons and his partners in Clarke, Chapman & Co., with the result that the former decided to terminate the partnership and to engage in manufacture on his own account. Now by the deed of partnership that had been entered into, it was agreed that if he withdrew from the partnership he was to be repaid the £20,000 he had originally invested in the Company, and, as regards the patents, it was agreed, as we have stated above, that any such patent taken out by a partner became the property of the partnership. The remaining partners at once offered to submit the matter of the, Parsons patents to arbitration, and transfer them to Parsons on such terms as the arbitrator should award. The arbitration took place, Gainsford Bruce, Q.C., M.P., being the arbitrator with counsel and experts on both sides. One side argued that the patents were of great value; the other that the measure of success so far attained was not sufficient to establish that such was the case. Besides, there were prior patents, which might or might not be held to invalidate Parsons's patents if tested in the Courts in the future. This was certainly true, and it may as well be stated that every type of steam turbine that has been brought out has been anticipated previously in the patent records. All that it has been and is possible to patent is some detail or other of construction. A further interesting fact may be mentioned and that is that there has never been a steam turbine patent action. Any one of the principal patents could have been assailed on the ground of anticipation if anyone had thought it worth while. Perhaps it has been the practice of granting licences that has saved the situation. How different in Watt's case! His was a master patent and unassailable except on legal points, yet he was badgered for years. Perhaps if he had granted licences, like Parsons and the rest did, Watt might have been let alone.

Before the arbitrator's award was made, however, Clarke, Chapman & Co. offered to retain the patents and continue on their own account. Parsons accepted these terms of settlement, thus relinquishing the axial flow turbine and the drum armature dynamo.

With courage and energy Parsons started afresh. If he could not build an axial flow turbine he could still build a radial inward or outward flow turbine, although he realized that it was inferior, as it proved, from a constructional point of view. With the aid of friends he started the firm of C. A. Parsons & Co. Ltd., at Heaton Works, Newcastle-upon-Tyne, subsequently to become world-famous.

This radial flow period was marked by the construction of the first condensing turbine, a unit of 100 kW., supplied to the Cambridge Electric Lighting Company in 1891. A very elaborate test upon it was carried out by Professor, afterwards Sir Alfred, Ewing, F.R.S., whereby it was demonstrated to all the world that it was more economical in steam consumption than the reciprocating engine of equal capacity of that time. Already favourably known for the small space occupied per kW., for silent running and reliability in service, the steam turbine now assumed the rôle of the world's prime mover.

The Cambridge outward flow turbo-alternator was of such importance that the particulars are worth recording:

Nature of current	Single phase, 80 cycles, 2000 volts.
Rated capacity	100 kW.
Number of wheels	7.
Revolutions per minute	4800.
Steam pressure	140 lb. per sq. in.
Temperature of steam (saturated)	361° F.
Vacuum at full load	26·5 in. Hg (Bar. 30 in.).
Steam consumption at full load	37 lb. per kWh.
Steam consumption at half load	39 lb. per kWh.
Overall length	14 ft.
Overall breadth	3 ft.
Total weight, without condenser	4 tons.

After long and honourable service it was bought in by the firm and presented, in 1925, to the Science Museum, where it has a prominent position close to the pioneer turbine of 1884.

It is well to realize why the condenser, important as it was in Watt's time, is more important still to the steam turbine, although for not quite the same reason. The explanation of this has been given so succinctly by Professor Andrade that we cannot do better than quote it:[1]

In a reciprocating engine it is impossible to take advantage of a very high vacuum in the condenser, for we cannot have a cylinder big enough to expand the steam to a very low pressure, nor wide enough passages from the cylinder to the condenser to let the very low pressure

[1] E. N. da C. Andrade, *Engines*, 1936, p. 174.

steam out in time. With the turbine, however, we have room to expand down to the lowest pressures, and we can make enormously wide passages from the low pressure end straight into the condenser, since they do not have to be closed and opened by valves, as in the reciprocating engine, but are always open. We can therefore take advantage of the highest possible vacuum....It can be calculated that, starting with a fixed high pressure, for each inch improvement of vacuum between 25 inches and 27 inches, there is a gain of 4 per cent. in efficiency; a further gain of 5 per cent. if the vacuum is improved to 28 inches, and a still further gain of 6 or 7 per cent. if it is increased to 29 inches.

The intriguing spectacle of reciprocating engines and steam turbines in the same power station was now to be seen, for example, at Portsmouth. A radial flow turbine of the kind just mentioned but of 150 kW. capacity, and a Yates and Thom reciprocating engine driving Ferranti alternators at 100 r.p.m., ran in parallel. Mr Ferranti tells the story of the official opening of the Station which was celebrated by a municipal banquet. This over, the aldermen, councillors and guests went along to see the Station. At that moment the alternators were on different circuits and were not absolutely in parallel, whence arose the stroboscopic effect of the Ferranti alternators appearing to run *backwards*. None of the banqueters dared confess to seeing this optical effect lest he should thereby stand self-accused of having dined "not wisely but too well".

We cannot linger over this period. Parsons himself, as time went on, felt more and more the limitations of the radial flow type. Clarke, Chapman & Co., during the same time, had done very little with the axial flow type; it was obvious that the patent would be of most use in the hands of the patentee. As a consequence, negotiations were opened, the partners were too level-headed to pursue a dog-in-the-manger policy, and eventually, in December 1893, Parsons bought in the patent for the modest sum of £1500.

Immediately a great impetus was given to development, not only in design but in size. A jump was made from 150 kW. to 350 kW., a size of unit then considered colossal. The consequential changes in design were numerous. Parsons decided that the blading should be of a uniform height throughout, viz. 1 in., and to allow for the increasing volume of the steam as it passed from stage to stage, the blading was accommodated on portions of the rotor of increasing diameter—in this case of five different diameters. This stepping of the rotor is a practice that has been followed ever since. To keep down the overall length and halve the blade-tip leakage, Parsons determined that the turbine should be of the single or unidirectional flow type. The consequence of this was that the end pressure, previously balanced internally, had now to be cared for. This was done by "dummy" pistons, one for each of the five different diameters of the rotor.

It should be mentioned that the set described was made for the Metropolitan Electric Supply Co., of London, for their Manchester Square Station. They were in an awkward dilemma because the Willans engines which had been installed had created such a nuisance to the surrounding residents by their noise and vibration, that an injunction against the Company had been granted, and the latter were ordered to abate the nuisance or to cease working altogether. The 350 kW. set was so satisfactory in working and so effective in abating the nuisance, that the situation was saved and more sets were at once ordered.

Nine years of the term of the patent had gone by and had in a sense been lost by its retention by Clarke, Chapman & Co.; the date of its expiration was looming in the distance and Parsons, who was a good business man, saw that there would not be time to secure any really adequate return before the patent would be open to public use. He therefore decided to apply for an extension; such an application in these days is comparatively rare, and it excited much interest, not altogether because of this, but because engineers and commercial men were now fully alive to the value of the patent. The appeal came before the House of Lords Judicial Committee of the Privy Council on April 19, 1898. During the appeal quite a large number of facts as to the technical progress made and the financial results of the firm were revealed. Suffice it to say that the judgment concluded as follows: "Their Lordships are of opinion that Mr Parsons has not been adequately remunerated and they will consequently make their report to that effect. They have already intimated that the patent should be prolonged for five years." The Privy Council granted the appeal—the extension to the patent was sealed on June 22, 1898, and thus it had a total existence of nineteen years terminating in 1903.

It is a matter for curious speculation to observe how long have been the periods for which patent privileges or extensions of them in connection with the steam engine have been granted in this country. The Marquis of Worcester's patent for his "water-commanding" engine, 1663 (assuming it was a steam engine), was for a term of ninety-nine years; Savery's patent for his fire engine, 1698, and its extension, together covered thirty-five years; Watt's patent for the separate condenser engine, 1769, and its extension, lasted for thirty-one years, and lastly, as we have seen, Parsons's patent for his steam turbine, 1884, was extended to nineteen years, a steadily decreasing tale of years. This latter extension was no small blow to Parsons's competitors, who had been expecting to be able to share the field with him, without the necessity of having to take out licences.

We must now record what turned out to be a fresh triumph in the historical development of the turbine—an event that created somewhat of a sensation in engineering circles and one that established the fame of Parsons's turbine on the Continent of Europe. This was the supply of two turbo-alternators for the City of Elberfeld, Germany. They were of 1000 kW. capacity, i.e. double that of any turbine hitherto constructed; not only so but they constituted a departure in design and construction for they were the first double-cylinder or tandem turbines made. The stipulations were most stringent, yet exhaustive tests carried out by Continental engineers of the highest standing gave results that surpassed the guarantees; we can only mention that the steam consumption at full load was 9·19 kg. (20·2 lb.) per kW. hour.[1]

[1] Full details are given in *The development of the Parsons steam turbine*, 1936, p. 42.

We may bring this chapter to a close with a few words on the important subject of turbine blading. Blading the first turbine was effected, as we have said above, by milling the spaces out of the solid at an angle of about 45° on the periphery of the rotor wheel and on the segments inside the cylinder. Obviously this left straight instead of what was desired, viz. curved, passages of gradually increasing width, which involved that the blade should be of crescent-shape section. On this point we cannot do better than again quote Dr Stoney: [1]

In the early machines the blades were machined on the circumference of solid rings of metal, and their form was conditioned largely by the limitations due to the methods of manufacture. Various alterations were made as time went on, but the necessities of manufacture remained a controlling influence in regard to the shape.

In 1896 Parsons made an important advance (Patent No. 8698) by devising independent blading made of drawn strip metal. A plain or serrated groove was turned in the rotor and in the cylinder; a piece of strip of the right blade length was inserted in the groove followed by a correctly shaped distance piece, expanded by caulking; this was continued till the groove was filled. Dr Stoney relates how

In the early days Parsons used to come into the drawing office with a scrap of paper on which were written particulars of the blading he had decided on for a certain machine. This was set out to scale, and modified if necessary according to his judgement. The staff were completely puzzled as to how he determined the proportions of the blading. It was not until turbines had been built for nearly 25 years that members of his staff evolved a sound theory of the steam turbine, and when Parsons was informed of this and had approved of it, he was content to leave the design of the blading largely in their hands, though he continued to exercise the closest supervision over the details and the general mechanical design.

Dr Stoney goes on to say that

The blade-shape and the spacing illustrated in his patent of 1896 proved practically perfect for turbines with the moderate velocity-ratios of 0·5 to 0·65 employed in those days and for many years afterwards.

[1] Memorial Lecture, *loc. cit.* p. 17.

Except for slight improvements in detail, such as sharper edges and better finish, both blade shape and spacing have remained virtually unchanged to the present day, in spite of numerous investigations and experiments carried out with the object of finding a superior shape. For higher velocity-ratios slight modifications have been found advisable, but for the conditions for which the blades were designed it may be said that the form and spacing were ideal. The prescience of Parsons seems all the more remarkable when it is remembered that in 1896 the present knowledge of aerofoil shapes and of aerodynamics was non-existent.

To increase the rigidity of long blades, one or more circumferential lacing strips are soldered into notches in the blades. Blading is a tedious hand operation and to speed it up the idea of making blades in segments was introduced in 1899 (Patent No. 16,288) by Parsons, Stoney and D. F. Fullagar. One way was to let the blades into notched strips of brass and bend over the teeth so left to hold the blades to the required curvature; another, known as the "rosary" method, was to thread the blades and distance pieces on a wire and build them up on a former. The resulting segments are caulked into the grooves in the same way as the individual blades were done.

So far we have concerned ourselves entirely with the work of Parsons and justly so, for he was the great pioneer of the steam turbine. His portrait (see Pl. VI) is taken from a photograph of date 1919 when he was sixty-five years of age, obligingly lent by Messrs C. A. Parsons & Co. The pose is characteristic of the man, as the author who knew him personally can testify.

To the other inventors who were pursuing parallel lines of experiment, we now turn.

CHAPTER XII

OTHER TURBINE PIONEERS

De Laval's experiments—His impulse turbine—Rateau's multi-stage turbine
—Curtis's velocity-compounded turbine—Ljungström's double-rotation
turbine—Stumpf's mixed turbine—Turbine blading.

PERHAPS we ought to have paused ere this to record
what other persons were doing in the steam turbine
field. In particular we now turn to the notable achievements of Carl Gustaf Patrik de Laval (1845–1913), who was
quite as early in point of time as Parsons in beginning experiments on the steam turbine. De Laval was descended from one
of those soldiers of Napoleon who went to the support of
Marshal J. P. J. Bernadotte, who later became Charles XIV,
King of Sweden and Norway. Of a highly inventive turn of
mind, de Laval had taken up the construction of separators of
cream from milk. Apart from other details he was faced with
the problem of attaining the high speed of rotation which is
necessary if such machines are to function successfully. With
the hand-driven type such speed is attained by means of worm
gear, but in the case of the power-driven machine de Laval
conceived the idea of getting the speed directly from a turbine.
He began, naturally enough, with the Heron type or steam
Barker's mill and designed an S-shaped rotating arm which
functioned fairly well, but for the reasons already given (see
p. 190) he gave it up and turned his attention to an apparatus
like Branca's (see Fig. 59), analogous to the Pelton wheel or
hydraulic impulse turbine, where a jet of water issuing from a
fixed nozzle strikes curved buckets fixed on the rim of a wheel
which consequently rotates at a high speed. In the middle of
the bucket there is a dividing wedge which splits the jet so that
the water is liberated on either side of the wheel. Theoretically
the water ought to be "dead" when it is liberated and to get
the best results the wheel ought to revolve at half the velocity

of the jet. In practice there are losses due to the water drops carried round on the wheel, friction, etc., so that the efficiency is usually about 85 per cent. It will be understood that it is unnecessary to apply the jets of water all round the circumference of the wheel. A single nozzle may be and is most frequently used.

Now substitute steam for water: as before, the same pressure exists on both sides of the wheel. The steam has to do work by changing its velocity. This means that the steam should be expanded down to the pressure of the exhaust before being applied to the wheel, so that the steam may issue from the nozzle with the velocity corresponding to the energy given out by expansion. A plain nozzle like that in the Pelton wheel will effect this if the ratio between the absolute pressure of the steam on the two sides is not greater than 0·58 for saturated steam or 0·54 for superheated steam, or to put it in another way the upstream pressure must not be more than twice the downstream pressure for the nozzle to be efficient. But de Laval wanted to employ much higher ratios than this and it was his great discovery in 1889 that he could employ very high ratios by making his nozzle with an expanding orifice (see Fig. 60), whereby the conversion of the pressure of the steam into kinetic energy is so complete that high efficiency is attainable. Now the velocity of the vanes or buckets should, as we have said above, be half the velocity of the steam, but here again the velocities involved with steam are much higher than those with water; for example, the velocity of steam expanded from a pressure of 200 lb. per sq. in. absolute to 29 in. vacuum is theoretically 4307 ft. per second. The peripheral speed of the wheel is less than half of this, i.e. approaching the speed of a rifle bullet, so that special design is necessary to render this possible. Perfect balance of the wheel, no matter how careful the workmanship, is unattainable, while in getting up to the speed involved, "critical" speeds must be passed which would give rise to destructive vibrations with rigid bearings. This was de Laval's greatest problem—he solved it by fixing his wheel in the middle of a long flexible shaft, one end of which is secured within a plain bearing while the other end is carried in a bearing spherically mounted. The wheel, when it gets up to speed, determines its own axis of rotation somewhat as does a spinning top. The speed is, however, far too high for ordinary machinery, so that de Laval reduced it by helical gearing in the ratio of 1 : 10, the countershaft being then at a speed suitable for driving one of his cream separators, or a

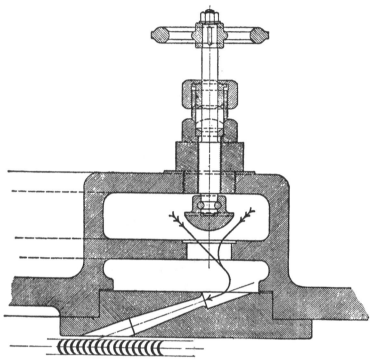

Fig. 60. Expanding nozzle, nozzle ring and stop valve, de Laval turbine.

Courtesy of Messrs Greenwood & Batley Ltd.

dynamo. The design of the wheel itself is equally refined. Wheels are of two patterns according to the size of the turbine. For those up to 150 h.p. the design is as shown in Fig. 61. The apparently excessive bossing of the wheel is because of the rapid rise in the tangential stresses as the centre is approached. In the larger wheels where the peripheral speed is still higher, it is safer not to have a bored-out boss at all, but to have flanges on the shaft attached by set screws (see Fig. 62). It will be noticed that grooves are turned under the rim of the wheel, so as to locate a fracture there should the wheel run away. Of course, the rim might cause damage, but nothing like that which would take place were the whole of the wheel to fly to pieces. The wheels are of drop-forged steel, and to accommodate the vanes or buckets holes are drilled transversely through the rim and saw-

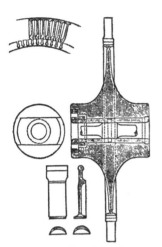 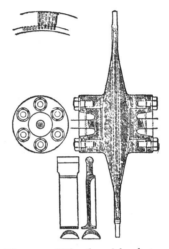

Fig. 61. Wheel and buckets used in small de Laval turbines.   Fig. 62. Wheel and buckets used in large de Laval turbines.

Courtesy of Messrs Greenwood & Batley Ltd.

cuts join them to the circumference, leaving keyhole slots into which the buckets are inserted sideways; the latter are crescent-shaped (see Fig. 61). The pitch of the buckets is about $\frac{1}{3}$ in., and flanges on the outer end are in contact thus preventing the escape of steam radially. To ensure that the steam will enter the bucket without shock, the nozzle is placed tangentially on the side of the wheel at an angle of 20°. The expanding part tapers at an angle of 10°, the portion next to the bucket being made cylindrical, as shown in Fig. 60. There are two or more nozzles to each wheel, and one or more may be shut off to meet a varying load without loss of efficiency. Governing is, however, effected by throttling.

The necessity for gearing is an objection; the construction of the turbine is such that it lends itself to small units, say up to about 300 h.p. Table II gives particulars of the principal sizes and speeds. The turbine is made in Sweden by Aktiebolaget de Lavals Ångturbin, Stockholm, and in England by Messrs Greenwood & Batley Ltd. of Leeds.

TABLE II. *Dimensions of de Laval Turbines*

Size of turbine		Mean diameter of wheel		Revs. per min.	Peripheral speed ft. per sec.
h.p.	kW.	in.	mm.		
1·5	1⅛	3	75	40,000	525
5	3¾	4	100	30,000	525
15	11¼	6	150	24,000	567
30	22½	9	225	20,000	774
50	37½	12	300	16,400	860
100	75	20	500	13,000	1135
300	225	30	750	10,600	1375

De Laval was a great pioneer and his portrait, taken from a photograph at the period of his inventions, now in the Science Museum, London, is deservedly included on Plate VI.

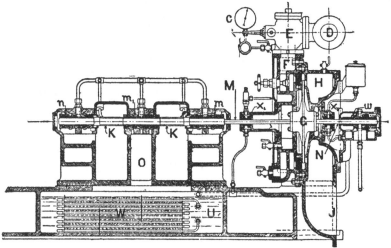

Fig. 63. Large de Laval turbine and reduction gear.
Courtesy of Messrs Greenwood & Batley Ltd

The two principles—impulse and reaction—that we have thus far exemplified are fundamental, but it will be realized that combinations of the two by compounding are possible, and these we must proceed to describe and illustrate. If we imagine the de Laval turbine pressure-compounded, that is, the steam expanded in successive stages in fixed guide blades, and the velocity due to each expansion absorbed by moving buckets on the rim of wheels alternating with guide diaphragms in series, we get the Rateau type of turbine. The speed of the rotor is thus reduced within practical limits, and the assemblage of these compartments or cells permits units of any size to be constructed. Such was the conception of Auguste Camille Edmond Rateau (1863–1930) of Paris, and to him is due the credit of having carried the idea into practice.

Rateau was born at Royan, Charente-Inférieure, France, October 13, 1863, and from boyhood he was attracted towards scientific studies. He became a pupil of the École Polytechnique, Paris, and came out first of his year in 1882. Following this he took the three years' course at the École Supérieure des Mines, and on passing out was appointed engineer in the Corps des Mines de l'État, stationed at Rodez. Subsequently he spent more than ten years in professorial chairs, proving himself to be a stimulating teacher. All the time that he could spare from his educational work he devoted to research and experimental work on the mechanics of fluids and their application in continuous flow to rotatory machines, whether they were for centrifugal fans, ventilators or steam turbines, in all of which fields Rateau made notable advances. We confine our attention, however, to the last named, which he took up in 1896. He patented his improvements in France in 1896–8, and in conjunction with G. E. Sautter and H. A. F. Harlé, of the firm of Sautter, Harlé & Cie, by whom his first machine was built in 1898, brought it into public prominence at the Paris Exposition Universelle of 1900. A portrait of Rateau from a photograph taken when he was 62 years of age is given on Plate VI by the courtesy of La Société Rateau.

As we have said, his turbine is of the pressure-compounded impulse or multi-stage type. At each stage there is a set of fixed nozzles in which the steam is expanded through a fraction of the total pressure drop, and there is a wheel which absorbs the kinetic energy thus generated. The steam passes through all the stages successively to the condenser. By thus expanding the steam in stages, the velocity it acquires at each stage is only that due to the pressure drop in that stage; hence each wheel has a correspondingly reduced rim velocity so that the turbine can be run slower, and yet efficiently, than if it were not so compounded. The first stage starts with "full admission", i.e. nozzles all round the circumference (for partial admission is inefficient), and the length of blade and height of nozzle increase from stage to stage onwards to meet the increasing volume of the steam as it passes successively through the stages.

The mechanical construction of the Rateau turbine is refined and ingenious. The general arrangement comprises, as we have said, a horizontal shaft on which are a series of wheels carrying buckets. Alternating with these wheels and fitting over the horizontal shaft are fixed diaphragms carrying nozzles. Thus each wheel is in a compartment by itself. The diaphragms are split on the horizontal diameter and are fixed in the cylinders, so that the turbine can be opened. Since the nozzles and buckets increase in size as the exhaust end is approached, to allow for the ever-increasing volume of the steam, this necessity is met by increasing the diameters of the rotor compartments. To simplify construction, instead of each compartment being of a different size, they are grouped. To avoid a long length of horizontal shaft without support, in large sizes the turbine may be divided into two parts, with an intermediate bearing between them.

There has been considerable development in the construction of the wheels and diaphragms. The first wheels were built up of stamped sheet steel riveted to a boss, and although light did not permit of a high peripheral speed. Heinrich Zoelly of Escher, Wyss & Co. of Zurich, Switzerland, who had been working also on the same type of turbine and built his first one in 1903, employed forged steel discs. Rateau followed his practice and forked the blades to straddle the rim, where they are held by rivets half in one blade and half in the adjoining one. The blades are milled out of solid nickel steel. There are holes in the webs of the discs to ensure equalization of pressure on both sides. The diaphragms are in halves; the top halves are secured to the cylinder and are bushed with anti-friction metal where they pass over the shaft. The bushes have labyrinth grooves turned in them to baffle the steam that may leak along the shaft from one compartment to the next.

La Société Rateau was founded by Rateau in 1903 with works at La Courneuve, Paris, and at Maysen, Belgium, to manufacture the turbine, and the many other apparatus due to his fertile genius. In 1908 the turbine was taken up in England under licence by Fraser & Chalmers Engineering Works, Erith, and by Metropolitan-Vickers Electrical Co. of Manchester. A typical machine built by La Société Rateau for an output of 8000 to 12,000 kW. is shown in Plate VIII. The upper half of the cylinder has been removed and the low-pressure rotor has been raised slightly out of its bearings. In this turbine steam is supplied at 426 lb. per sq. in.; the total temperature is 752° F. and the vacuum 29 in. of mercury (barometer 30 in.). The high-pressure rotor runs at 7500 r.p.m. and is geared to the low-pressure rotor, whose speed is 3000 r.p.m.

The disadvantage of the de Laval turbine is, as has been mentioned above, the very high speed of the buckets. The lowest number of revolutions being not less than, as will be seen from Table II, 10,600 per min. In order to reduce this speed resort has been made to the method of velocity-compounding. The steam, leaving the nozzle at a velocity much too great to be dealt with efficiently by a single row of moving blades at a reasonable number of revolutions per minute, is made to act upon two, three or (in marine practice) even four rows of moving blades on the same wheel, being reversed in direction by guide blades between the successive rows, each row of moving blades abstracting a portion of the velocity of the steam. The sole purpose of the guide blades is to turn it back to the next row of moving blades. All the wheels are mounted on the same shaft, and the velocity is thus brought down to a practical limit, say 350 to 450 ft. per sec. The advantage of this method lies in the ease with which low bucket speeds can be obtained with a small number of

PLATE VIII. RATEAU DOUBLE-CYLINDER GEARED TURBO-ALTERNATOR

Courtesy of La Société Rateau, Paris

wheels; the disadvantage lies in the high friction losses in the steam which render the possible efficiency materially less than with other methods of compounding. A practical drawback is that although the number of elements is small they are somewhat massive, as the buckets have considerable axial and radial depth.

The earliest velocity-compounded turbine was that of Charles Gordon Curtis of New York. He was born at Boston, Mass., on April 20, 1860, and for eight years practised as a patent lawyer. His interests were, however, in manufacture. He organized the Curtis Electric Manufacturing Co., of which he was President, to make motors, fans, etc., and thus his mind was directed towards the steam turbine, then in its infancy. Curtis's earliest patents were taken out in this country in 1896 (Nos. 19,246–8) and were for pressure-compounding, but he soon gave this up for velocity-compounding. The rights of his invention were taken up in 1897 by the General Electric Company in return for plant facilities and a staff for development at Schenectady, N.Y., where experimenting began. Serious problems of design and operation were encountered and after two years of intensive work in close touch with Mr Curtis, the engineer who was in charge of development recommended that tests be abandoned. Mr E. Wilbur Rice, Vice-President in charge of engineering and manufacture, who had sponsored the original contract with Curtis, unwilling to give in, invited Mr William Le Roy Emmet to investigate. We quote from the account he afterwards gave of what took place:

He (Rice) showed me reports of other engineers who had considered the experimental showing not good enough and recommended that the work be dropped. He asked me to look into it and to give him my opinion. I then met Mr C. G. Curtis for the first time and went carefully over his designs and constructions and analyzed his test results to the best of my ability....Mr Curtis gave me very clear and intelligent explanations of what he had done, and of his reasons, and I began to take a keen interest in the subject. I was very much alive to the great value of such developments in the electrical industry if they could be made successful. I reported to Mr Rice that while the actual steam economies were not as good as those of good steam engines, I thought that the work was important....I recommended that we go on with the work.

This advice was accepted and the first activity was to design a 500 kW. turbo-generator of the horizontal-shaft type for the Schenectady Works power plant. This took about a year, and its success was such that larger operations were planned.[1]

Fig. 64 is of historical interest as it shows the first 5000 kW. turbine built by the firm. This historic turbine developed the power stated at 500 r.p.m. with dry saturated steam of 150 lb.

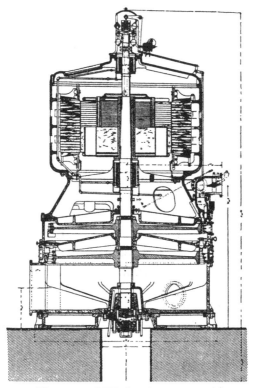

Fig. 64. First 5000 kW. Curtis steam turbine, 1903.
Courtesy of the General Electric Company, Schenectady, N.Y.

per sq. in. pressure, temperature 380° F. and exhausting to 28 in. vacuum. It will be remarked that the shaft was vertical as was then the practice. Advantages were claimed for this, but they were offset by more numerous disadvantages such as that the whole of the weight of the rotating parts was taken by a single footstep bearing, and that the condenser was inaccessible;

[1] Information kindly supplied by the General Electric Company, Schenectady.

the vertical-shaft turbine was soon abandoned and no other firm ever built one. The horizontal position is now, as with every other maker, stereotyped.

There were a number of nozzles depending upon the diameter of the wheel, each under the control of a drop valve actuated by a governor which shut them off in succession as the load was reduced. The nozzles were of the de Laval divergent type. Each turbine wheel ran in its own chamber and the steam passed from chamber to chamber by means of nozzles in the diaphragms separating the chambers. The succeeding wheels and chambers were closely similar, except that the nozzles and buckets successively occupied a larger portion of the periphery of the wheel to allow for the increasing volume of the steam, till at the last wheel they occupied the whole periphery. Fig. 65 shows the shape of the corresponding fixed and moving buckets. The latter are machined from the solid in a special form of shaping machine, except in the larger sizes, when a number have been cast together in bronze in segments and bolted in position on the wheel. The clearance between the fixed and moving buckets varied according to the size of the unit, from 0·02 to 0·08 in. The footstep bearing, on account of the enormous weight

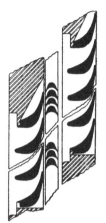

Fig. 65. Development of fixed and moving buckets of Curtis steam turbine.

Courtesy of the General Electric Company, Schenectady, N.Y.

it had to support, was of special design. It consisted of two cast-iron discs, the lower one being fixed and the upper one rotating with the shaft. The lubricant was oil forced between the discs at a pressure of 400 lb. per sq. in., which was just sufficient to keep the surfaces separate. The turbine was taken up in 1900 by the Allgemeine Electricitäts Gesellschaft of Berlin, and in 1903 by the British Thomson-Houston Company of Rugby, firms associated with the General Electric Co.

Our portrait of Curtis (see Plate VI) is taken from a photograph, dated 1923, when he was sixty-three years of age, obligingly furnished by the General Electric Co.

We have now to consider an entirely different and extremely ingenious arrangement of the steam turbine, the invention of Birger Ljungström. He was born on June 4, 1872, at Uddevalla, on the south-east coast of Sweden, but passed his boyhood at Stockholm, where his father was an engineer and surveyor. The young man was first attracted to the subject of steam turbines by seeing the turbine of de Laval, the expanding nozzle, flexible shaft and helical gearing of which called forth his admiration. In 1904 Ljungström made a sketch of a double-rotation radial flow turbine in which the de Laval nozzle was embodied. The idea of double rotation was not altogether a new one—it had been suggested in 1879. It involved the innovation that what had been the fixed blading should now rotate in the opposite direction to the moving blading at the same rotational speed and that work should be done with each half of the blade system. The advantage of this arrangement was that the turbine could be equivalent to an ordinary turbine running at twice the speed; in other words, the same efficiency as previously could be attained with a quarter of the number of stages, or a higher efficiency could be attainable if the number of stages was kept the same as before. The shafts, though not coupled in any way, could be kept uniform in speed by the generators which they were to drive being in parallel and mutually balancing. To carry out the idea the blading must be on overhanging discs facing each other. Ljungström made some preliminary experiments with such discs, then a novel idea, and found that the discs by gyroscopic action gave excellent results, proving that no difficulties would be encountered in running such wheels up to and through their critical speeds. His attention was, however, diverted by other experimental work. In the meantime he heard about the Parsons steam turbine, but it was not till 1901, when on a visit to Newcastle-upon-Tyne, that he actually saw one. At a certain college which we shall not name he saw a demonstration with a Parsons turbine of 30 h.p. The demonstrator was evidently not impressed with the machine, for he said it "was rather leaky" because "as much steam would run through when the turbine was standing still as when it was running"; this is instanced to show how little the steam turbine was generally

PLATE IX. LJUNGSTRÖM DOUBLE-ROTATION STEAM TURBINE UNDER TEST, 1910

Courtesy of the Inventor who is seen in the background

understood at the time. Ljungström, however, was not misled by this demonstration, because fortunately he was able to see over the Forth Banks Power Station of the Newcastle and District Electric Lighting Co., where he saw several Parsons turbo-alternators supplying single-phase current. His favourable opinion was further strengthened by reading of the tests of the Elberfeld turbine already referred to (p. 204). His interest in steam turbines was not awakened in real earnest until early in 1906, when he found time, owing to a temporary withdrawal by illness from everyday work, to mature his ideas and convince himself of his ability to solve the large number of mechanical and manufacturing problems involved in the practical realization of the new type of turbine. He worked on the problem from inside outwards, as it were, by starting with the manufacture of a single-blade ring of about 900 mm. diameter which he tested at a speed of 3000–4000 r.p.m. Further turbine designs were brought out, and on the strength of the results he got out drawings for a complete turbine section and submitted them to Professor A. Stodola of Zurich, who gave a favourable opinion as to the possibilities of the new turbine. Ljungström's first British patent was taken out in 1907 (No. 7833). Two years had been taken up with this experimental work. At the beginning of 1908, he succeeded in floating a company, the Aktiebolaget Ljungströms Ångturbin (ALA) of Stockholm, and to do so he had to enlist the support of persons who did not understand anything about steam turbines, after having tried in vain persons who thought they understood all about them !

At the beginning of 1910 the first turbine—a 500 h.p. set capable of overload to 1000 h.p.—was completed. The number of rings was 36, varying from 120 mm. to 500 mm. diameter. Although embodying about twenty improvements in detail, it ran straight away at full load at 3000 r.p.m. for several hours and showed a thermodynamic efficiency of 68 per cent. It is this turbine under test which is shown in Plate IX. The water-brake dynamometers at the ends of both shafts can be clearly seen. Ljungström is seen standing behind the plant.

By the beginning of 1912 a 1000 kW. turbo-alternator had been built and tested with still better results, i.e. an efficiency of 77 per cent. This machine was bought for the Willesden

Power Station, where it ran on tramway work for twelve years. The performance of the turbine now brought it to the front rank. In 1913, another company, the Svenska Turbinfabriks-Aktiebolaget Ljungström (STAL) of Finspong, Sweden, was formed to manufacture the turbine, and in the same year it was taken up by the Brush Electrical Engineering Co., of Loughborough, England. The Great War, however, caused a gap in the commercial exploitation of the turbine.

The inventor was greatly assisted in the development of the turbine, subsequently to the initial stage, by his brother Fredrik (b. 1875) and by Ing. O. Wiberg. In 1916 Ljungström relinquished his position with ALA and in 1923 that with STAL, in order that he might be free to devote himself anew to technical work; this, however, has been in another direction altogether. Our portrait of Ljungström (see Plate VI) is taken from a photograph[1] of the time when he had completed his invention, that was in 1914, when he was forty-two years of age.

It is now proper to bring out somewhat more clearly the principle of the turbine[2] and show the flow of the steam, points which will be readily grasped from the elementary diagram, Fig. 66. The blade rings interpenetrate or interleave and at each interleaving the flow of the steam is reversed in direction. Steam passes through an emergency valve and governor and enters through the pipe P into the steam chests CC. Steam then passes through the holes HH in the hubs of the disc into the centre of the turbine, whence the steam flows radially outwards between the concentric rings BB, mounted on overhung discs DD, and to the shafts of these the generators GG are keyed. Steam exhausts into the low-pressure casing and thence to the condenser, which is underneath the turbine. KK are the axial shaft packings, LM the labyrinth packings limiting the leakage of high-pressure steam and balancing the axial pressure tending to force the discs apart. The condenser forms the foundation of the turbine, while the casings and the stator frames make up a rigid tubular structure.

[1] The photograph has been kindly lent by the inventor himself.

[2] Cf. A valuable paper by P. S. Wakefield, M.Sc., read before the Electrical Power Engineers Association, November 30, 1937, and March 14, 1938.

With this explanation it is now possible to grasp the refinements shown in Fig. 67, which shows (due to consideration of space) the top half only of the blade system, in this case of an 1875 kW. turbine. It will be observed that at the exhaust end single blade rings have been replaced by two rings in parallel; this is to deal with the increased volume of steam; hh and h_1h_1

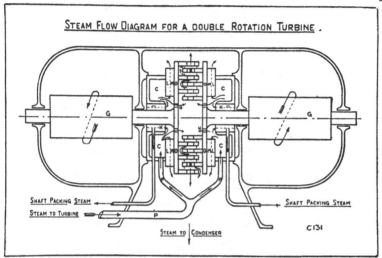

Fig. 66. Diagram of steam flow of Ljungström double-rotation steam turbine.

Courtesy of the Brush Electrical Engineering Co. Ltd.

are the discs, ff the labyrinth steam packing (see Fig. 68), ww the shaft packing (see Fig. 69) and ee are the steam chests. A clever feature of the Ljungström design is the use of expansion rings of dumb-bell section, which will be seen between all connected parts subject to temperature changes, to enable them to expand and contract freely. One of these rings will be seen clearly on Fig. 70 which is a section of a single blade ring; 1 is the blade disc and 4 the expansion ring which is held by rolling over the material at 5. It is this that enables the largest unit to be run up to speed with impunity in a few minutes from cold.

The demand for larger outputs and higher vacua has led to the addition to the turbine of axial flow exhaust stages, to which the position of the working parts readily lends itself. This addition was discussed with Professor Stodola as early as 1910 but was not embodied till later. High efficiencies have been

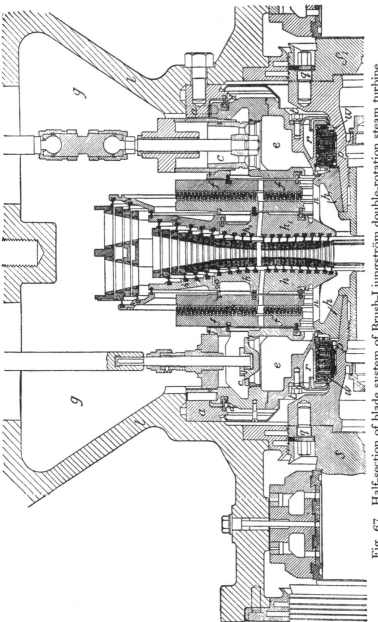

Fig. 67.   Half-section of blade system of Brush-Ljungström double-rotation steam turbine.

Courtesy of the Brush Electrical Engineering Co. Ltd.

obtained. A 50,000 kW. set supplied by STAL to the Swedish Water Power Board in 1932 showed a steam consumption of 8·8 lb. per kWh., corresponding to an efficiency of 91 per cent

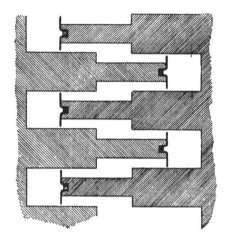

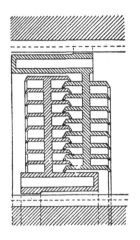

Fig. 68. Section of Brush-Ljungström labyrinth disc packing.

Fig. 69. Half-section of Brush-Ljungström labyrinth shaft packing.

Courtesy of the Brush Electrical Engineering Co. Ltd.

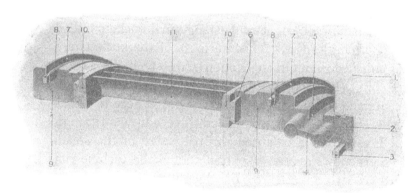

Fig. 70. Section of Brush-Ljungström single blade ring.

Courtesy of the Brush Electrical Engineering Co. Ltd.

relative to the heat drop after passing the throttle valve. The largest set so far made in Great Britain, a 37,500 kW. set, the first of two, was installed last November in the Southwick Power

Station of Brighton Corporation. Spot tests have given a steam consumption as low as 8·4 lb. per kWh. The limit of size does not yet appear to have been reached.

The possibilities of velocity compounding are not, however, exhausted. There is the ingenious idea of doing so in a single set of buckets which, actually suggested as long ago as 1863, was not reduced to practice until 1903 by Professor J. Stumpf of Berlin. The steam after leaving the nozzle enters the wheel buckets, gives up some of its kinetic energy and is discharged in the reverse direction. It is then caught up by the first of four semicircular guide chambers and returned to the wheel buckets, where it gives up more of its energy, and so on through the successive guides. Steam may be passed, for example, five times through the wheel, and the resulting bucket speed is about one-fifth of that required with a single nozzle. An additional guide bucket is added behind the nozzle to catch any steam that may miss the first bucket; the efficiency of the turbine is thereby considerably improved. The arrangement does not permit a high economy but meets the demand for a simple and reliable turbine for driving small machinery of low power requiring little attention.

In an example developing 5 h.p. the rotor is a steel forging 12 in. diameter having 40 semicircular tangential buckets milled in the rim at an angle of 33°. The surfaces of the buckets are ground and polished to minimize friction of the steam. The rotor is mounted on a horizontal shaft in ring-oiled bearings on spherical seats and fitted with metallic packing where it passes through the casing. A spring-controlled governor acts through a bellcrank lever upon a balanced throttle valve in the steam pipe. An emergency governor is fitted should the turbine run away. There are two nozzles, one of which is fitted with a stop valve to take care of partial load without loss of efficiency. The speed is 3000 r.p.m. It is possible to have more nozzles and thus increase the rating of the unit.

It will be realized that further combinations in turbines are possible; in fact, nowadays, scarcely any turbines are of one single type.

CHAPTER XIII

GENERAL DEVELOPMENT FROM 1900
TO THE PRESENT TIME

Chart of progress—Alternating current generators—Newcastle-upon-Tyne Electric Supply Co.—Parsons's "augmentor"—25,000 kilowatt generator for Chicago—The Great War—Increase in size of units—Turbine blading —Conclusion.

I N the preceding chapter the development of the principal types of turbine has been traced and we can turn usefully to the broad consideration of some of the outstanding events from 1900, which was the turning-point when turbines began to be employed in preference to reciprocating engines, down to the present day.

Progress has been phenomenal and this will be seen epitomized in the chart to be found on p. 226. The chart embodies the experience of a leading firm of American turbine-makers— the Westinghouse Electric & Manufacturing Co., originally the Westinghouse Machine Co., who were the earliest to become licensees of Parsons; this was in 1895 and they have been co-workers for forty years. Although the progress charted is that of one firm only, it may be accepted as typical.

The chart will repay close study. The first item that arrests attention is the rise in steam pressure and the increased vacuum that has been attained. The resulting increased internal efficiency is reflected in the diagram of total temperature at the throttle valve, and in the B.Th.U.'s available in the heat cycle. As regards vacuum, we can say that it is not usual to provide a higher vacuum than 29 in. of mercury with a cooling water temperature of 55° F., so that the direction of further progress is obviously towards higher pressures.

The steam consumption is not charted, but it should be noted that while in 1900 in the Elberfeld turbine (see p. 204) the consumption was 20·2 lb. per kWh. at full load, to-day in a modern turbine it is less than one-half of this (see p. 224). As

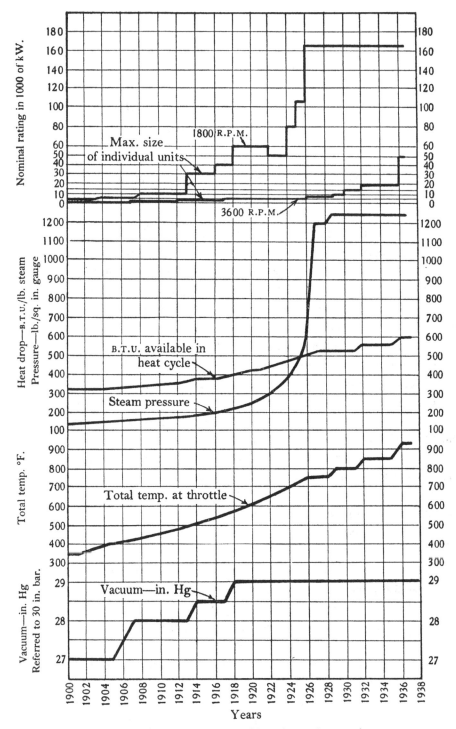

Fig. 71. Chart of progress in steam turbine size and operation, 1900–38.

Courtesy of the Westinghouse Electric & Manufacturing Co., Pittsburgh.

if to defeat this effort of the engineer, however, the cost of fuel has advanced to such an extent as almost to cheat him of the fruits of his labours.

The increase in size of units is very marked, especially in the United States—this leads also in the direction of efficiency. With regard to the space occupied by plant—an important point to the user—it may be stated that a kW. can be generated to-day in a space only one-tenth of that required in 1900 for that duty.

The low-grade heat rejected in the condenser circulating water is a loss we would fain avoid. An attempt has been made with some success to use it for horticulture.

We can now consider in some detail certain other aspects of the period under review. Prior to 1900 all turbo-generators were of the direct current, or else of the single-phase alternating current, type. In this year the first, as far as we can learn, three-phase generator, the type now used exclusively, was built. It was small—150 kW. at 2520 r.p.m.—and was not made for a power station but for a colliery, that of Lord Masham at Ackton Hall, Featherstone, Yorks, by C. A. Parsons & Co.

In 1900 the Newcastle-upon-Tyne Electric Supply Co. inaugurated in England the public supply of three-phase current. The voltage was 6600 and the cycles 40 per sec. The generators chosen were driven by triple expansion four-cylinder reciprocating engines of the inverted type (see p. 152), then the most efficient prime movers available. Shortly after commencing supply, additional plant became necessary, and as showing how the reciprocating engine was being forced out of the field, the consulting engineers, Messrs Charles Merz and William McLellan, decided upon the turbo-generator; the machine was larger than any previously made, since it developed 1500 kW. at 1200 r.p.m. It was a single-cylinder machine and the rotor was in three steps, increasing in diameter successively in the ratio $\sqrt{2}:1$, a proportion that has been adhered to at the Heaton Works over many years. That the adoption of the turbo-generator was justified is shown by comparative trials of it and the reciprocating engine carried out at Neptune Bank Station in 1904. A summary of the results is as follows:[1]

TABLE III. *Comparative tests of reciprocating and turbine machinery*, 1904

Reciprocating engine

Test no.	Speed r.p.m.	Average load, kW.	Steam press. at stop valve, lb./in.^{2}	Temp. at stop valve, deg. F.	Average vacuum, in.	Steam consump., lb./kWh.
1	100·8	798	203	482	23	24·22
2	87·9	589	204	476	24·25	23·48
3	74·5	316	201	457	25·6	23·42
4	52·5	96·5	204	425	24·5	30·93

Steam turbine

1	1183	1478	203	453	25·9	18·67
2	1110	1092	204	448	26·7	19·65
3	903	600	203	440	26·8	24·01
4	649	196	212	436	27·7	35·01
5	1210	1822	203	480	25·5	17·80

It will be observed that the steam consumption of the turbine at full load was about three-quarters of that of the reciprocator, while at low loads the latter was more economical. Of course the latter consideration carried no weight as the plant was only intended to work at full load. A further point of note was that the steam consumption of the reciprocator had increased since being set to work in 1900 by 3·3 lb. per kWh., whereas the consumption of the turbine was within a fraction of that observed when first installed a year earlier.

It should be mentioned that these tests were carried out at the request of a commission of leading marine engineers and shipbuilders appointed by the Cunard Company in 1904 to advise as to the type of machinery to be fitted in the transatlantic liners "Lusitania" and "Mauretania" then being built. With these results before them the Commission could arrive at

[1] *The development of the Parsons steam turbine*, 1936, p. 65.

no other conclusion than to recommend the adoption of turbine machinery. In a further test—that of endurance—one of the turbines ran for 7500 hours during which it was stopped for a total of 52 hours only and then merely for inspections.

We have already stressed the importance of the vacuum in a steam turbine and it was the realization of this that led Parsons to the invention of the vacuum "augmentor" or "intensifier" to assist the air pumps. This, which he had patented in 1902 (No. 840), is on the principle of the steam jet air pump and it was at Carville that it was used for the first time.

Space does not permit of telling the full story of the turbine, but its march has been that of a conqueror. Astonishing as had been the successive advances in size of turbines and the rapidity with which they followed one another, no single step was so great as that taken in 1912 when the construction of a machine four times the output, viz. 25,000 kW., of any existing machine was undertaken by Parsons to meet the wishes of the Commonwealth Edison Co. of Chicago for their Fisk Street Power Station. It was not only in output that the turbine was to surpass all previous efforts, but also in steam consumption, which Parsons guaranteed should not exceed $11\frac{1}{4}$ lb. per kWh. The design embodied the new feature of two cylinders in tandem, the low-pressure one being double-flow, an arrangement now adopted almost universally in large axial flow turbines. We cannot go into the details of the construction of this, at the time, gigantic machine, but an idea of its size can be gathered from the fact that its overall length is 76 ft. 2 in., its width 21 ft. 8 in. and its height 30 ft., 20 ft. of which is below engine-room floor level. The exhaust blading is 19 in. long; had it not been for the adoption of the double-flow principle, it would have had to have been twice as long.

The turbine fulfilled anticipations. Acceptance tests were carried out in 1914 and were repeated four years later when previous experience was remarkably amplified by finding that the steam consumption was slightly less than when the turbine was new. In other words a well-designed reaction turbo-generator suffers no appreciable deterioration in efficiency after protracted use. No wonder then that the machine is still in service and that it has earned from the station engineers the soubriquet of "Old Reliability".

Then the Great War intervened and development was checked. Power stations and factories had still to be supplied with plant, but it had to be of standard design that could be turned out quickly.

After the War, development began again[1] and in nothing has this been more striking than in the size of the unit. We may cite the 168,000 kW. reaction set for Hell Gate Station, New York, 1928, by Brown-Boveri & Co., the 165,000 kW. four-cylinder impulse-reaction set by American Westinghouse for the same station, 1928, and the colossal 208,000 kW. impulse set for the State Line Generating Station, Illinois, 1929. Such capacities have not been exceeded and there seems to be no need to do so. The tendency is to obtain the maximum possible output at moderate speeds; this is conditioned by the peripheral speed of the last row of moving blades. In Great Britain this has been kept down to about 1000 ft. per sec., but in the United States as much as 1200 ft. per sec. is allowed.

As representative of modern practice we have selected for illustration a Parsons two-cylinder tandem turbine of pure re-action type with double exhaust designed for an output of 30,000 kW. at a speed of 3000 r.p.m. (see Fig. 72). It is designed to operate with steam at 400 lb. per sq. in. gauge pressure, superheated to 800° F. The steam expands in the high-pressure cylinder to about 12 lb. per sq. in. absolute, enters the low-pressure cylinder at the centre of its length and exhausts at both ends to the condenser. The steam chest (not shown) contains two double-beat governor valves controlled by oil relays, and a separate emergency valve that shuts off steam automatically should the machine exceed a pre-determined speed. The blading is of the "end-tightened" construction, whereby large radial clearance is obtained with adjustable axial clearance. The con-

---

[1] Cf. Gibb, C. D., "Post-war land turbine development", *Proc. Inst. Mech. Eng.* 1931, p. 413.

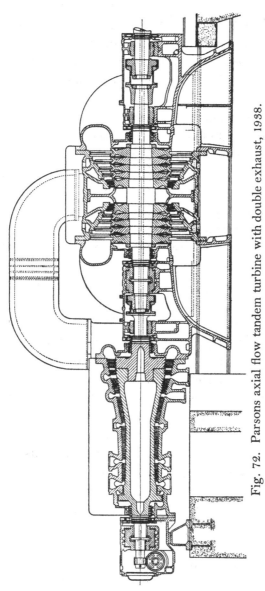

Fig. 72. Parsons axial flow tandem turbine with double exhaust, 1938.

Courtesy of Messrs C. A. Parsons & Co.

struction of the rotor is noteworthy. The high-pressure body is a hollow forged drum and the thickness of its walls approximates to that of the enclosing cylinder, so that both it and the rotor expand equally and consequently working clearances remain unaltered. The low-pressure part of the rotor is solid forged. It is stepped to accommodate the forged turbine discs which are shrunk on. These discs are butted together at both hubs and rims so that vibration is eliminated.

Another turbine representing modern practice is shown in Fig. 73. It generates 4000 kW. at 60 cycles, 3 phase and 480 volts ; it operates between 400 lb. pressure, 165° F. superheat and 2 in. Hg absolute pressure. It will be observed that it is a double-extraction turbine, steam being "bled" at 150 lb. and 40 lb. pressure respectively.

Research to find out the best materials for every part of a turbine has engaged attention. Rotor forgings are made of alloy steel to stringent specifications. Cylinders are made of steel for all temperatures above 450° F. when cast iron begins to be subject to growth and distortion. In the turbine, Fig. 72, the cylinder and steam chest are made of molybdenum steel. For blading, stainless iron or steel is largely employed because this material is resistant to corrosion by oxygen and carbon dioxide from which steam is never quite free. Such material, however, does not withstand so well the erosive action of the water particles entrained with the steam. Erosion is encountered at the exhaust end of the turbine where the steam is very wet. It occurs in the last row of moving blades, mainly at their outer ends where the comparatively slow-moving droplets of water are struck by the blades with the greatest force. To protect the blading against erosion, a hardened tungsten steel shield is brazed along the inlet edge of each blade on the last moving row.[1] The integral blade, i.e. one rolled, or forged solid and milled out, including a serrated root to fit into or over a groove on the turbine disc, is now universally adopted, particularly for the exhaust end where the blades are at their maximum length and the stresses correspondingly high. Blading of this kind is shown

[1] Gibb, C. D., *loc. cit.* p. 435.

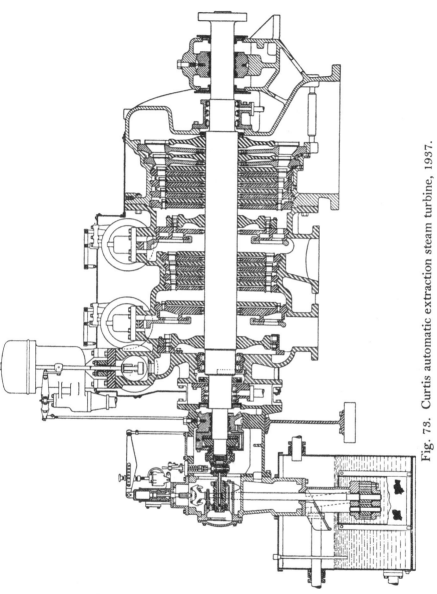

Fig. 73. Curtis automatic extraction steam turbine, 1937.
Courtesy of the General Electric Company, Schenectady, N.Y.

in Fig. 74. Integral blading has proved successful and a case of failure is rare.

Stripping of blading, owing to expansion or distortion causing the rotor blades to touch the cylinder or *vice versa*, caused trouble at an early date. The difficulty was overcome by thinning the blades at the tips (Parsons's and Stoney's Patent No. 22,217 of 1905) so that they simply wear away where they touch, if they happen to do so, instead of being stripped bodily.

Mention has been made of labyrinth packing. Although the corresponding problem of preventing the leakage of steam along a reciprocating rod had been solved (see Fig. 45), the comparable problem of preventing leakage past a rotating shaft, a problem that did not arise till the advent of the steam turbine, required a different solution. This lies in opposing to the passage of the steam a multiplication of small baffles, i.e. giving the steam

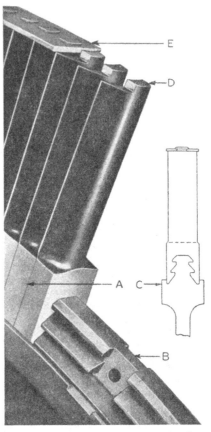

Fig. 74. Method of inserting buckets with outside dovetail on wheel of Curtis steam turbine.

Courtesy of the General Electric Company, Schenectady, N.Y.

a kind of hurdle race. Parsons's first method was to turn V-shaped projections on the rotor where it passed through the cylinder and run white metal into the projections. In 1896 this was elaborated: brass strips or rings in segments fixed on the cylinder projected into and touched the sides of grooves cut in the rotor. If slight axial or radial displacement takes place the strips or rings wear away without causing damage. The form

which this now universal device takes in the case of the Ljungström turbine is shown in Fig. 68 for disc packing and in
Fig. 69 for rotor packing.

More could be said but we must make an end. Reviewing
critically from many angles the progress that has been made
during the period covered by this chapter, it is impossible to
avoid the conclusion that progress, phenomenal at first, has
become slower and slower as years have gone by. It seems as
if the limit of economic size and speed of the turbo-generator
has been reached and with the materials at command much
higher temperature (and as a consequence efficiency) is not
attainable. We look forward with hope to the advent of another
Newcomen, Watt or Parsons to blaze a new trail.

CHAPTER XIV

DEVELOPMENT IN BOILERS, 1901 TO THE PRESENT TIME

Natural circulation boilers—High-head boilers—Forced-flow circulation—Indirect evaporation—Rotating furnaces—Pressure combustion.

IN the preceding chapter the view was expressed that the development of the turbo-generator was coming to a standstill, but perhaps this is too dogmatic an opinion seeing that the progress of the boiler which reacts continuously on that of the generator has, during the period under review, been considerable. Perhaps it would be safest to say that the baton in the relay race is at present with the boiler. Such progress has been not only along orthodox natural circulation lines, but there has been also markedly intensive research in forced-flow directions. These advances are being made concurrently and may be said to have in view the same object, namely, to increase heat output from a given weight of boiler.

To take first the directions in which the orthodox natural circulation boiler has been improved, we may mention that much research has taken place on combustion and on heat transfer. The importance of ensuring complete combustion of the fuel so that the resulting gases are incandescent before they reach the boiler surfaces has been realized, and this is effected by providing a combustion space of ample size. In the case of solid fuel, now almost universally burnt upon a chain-grate stoker, the combustion chamber has displaced the fire-brick arch over the grate which was formerly used to drive off the volatile matter in the fuel. A large combustion chamber is still more important when, in order to effect still higher heat releases, the coal is pulverized or the fuel is in liquid form, not infrequently in conjunction with pre-heated air. In the case of pulverized fuel the temperature may be allowed to become so high that the ash attains the molten condition and can be run off as slag at a tap-hole. Incidentally, the discharge of dust from the

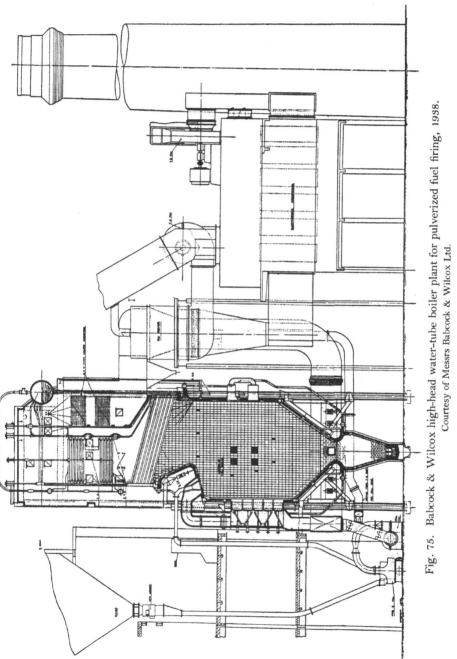

Fig. 75. Babcock & Wilcox high-head water-tube boiler plant for pulverized fuel firing, 1938.
Courtesy of Messrs Babcock & Wilcox Ltd.

chimney—an intolerable nuisance in residential areas—is obviated. To meet these high heat releases, tubular water walls such as are employed in the Bailey furnace, have been introduced.

When high furnace temperatures were first mooted, it was argued that natural circulation would be inadequate because the specific steam volume decreases as the pressure increases. Great attention has been directed to this point but without giving much help to the practical man, except to prove that such fears are unfounded. The boiler surface relative to that of the economizers and superheaters has been diminished and the present tendency is in the direction of the high-head boiler, where the vertical distance between the drum and the convection surface is longer than hitherto and this head therefore helps to overcome circulation defects. The boiler of this type that we illustrate (Fig. 75) is supplied by Babcock & Wilcox Ltd. It is a high-head controlled gas flow boiler fired by low volatile pulverized fuel, to deal with which it has a Bailey hopper bottom furnace, i.e. with water-cooled walls. It will be remarked that these boiler plants comprise in themselves: pulverizers, air preheaters, superheaters and feed-water heaters.

While there has been previously undreamt-of rise in pressure and in output of boilers of the orthodox design, there has been much experimental work going on along novel lines, in both cases perhaps as much for marine as for land purposes. For the latter the directions in which experimental work has taken place[1] are as follows:

1. Forced flow.
2. Indirect evaporation.
3. Rotating evaporation furnaces.
4. Pressure combustion.
5. Other working fluids besides water.

This is not a strict classification as boilers under one heading will be found to possess characteristics of those in other classes.

[1] For a detailed summary of the present position of such boilers, to which the author expresses his acknowledgment, see Dagobert W. Rudorff, *Steam Generators*, 1938.

The forced-flow principle may be sub-divided according as (a) the flow is once-through or (b) the circulation is continuous. In (a) the boiler surface consists of a long continuous coil, or else of short coils in parallel, feed water is forced in, preheated, evaporated and the steam evolved superheated. Due to the great length of the coil or circuit the pressure drop and the power required by the circulation feed pump is considerable. Owing to the small amount of storage space the feed, the fuel and the air supply have all to be under instantaneous and interlinked control.

Of this type is the boiler designed in 1922 by Mark Benson. His idea was to generate steam at the critical pressure (i.e. 3226 lb. per sq. in. and 706° F.), where the specific volume of water and that of the steam obtained from it are identical. He believed that at any lower pressure the bubbles of steam would occupy so much of the tube surface that they would blanket off or slow down the transmission of heat between the tube wall and the water. He recognized the fact that with water as the working fluid at the critical pressure no higher efficiency can be attained.

An experimental boiler was built by the English Electric Co., Ltd., at Rugby, and it was found possible to operate it at the critical pressure. Great difficulty was caused, however, by burning of the tubes at the point of greatest heat transmission, and it was not until after some years had passed that it was found that this was caused by the deposit of salts at this point. This was not found out earlier because by the time the boiler was laid off, the salts had redissolved and nothing was to be found. This deposition can be overcome by periodical washing with boiling water. However, the advantages of critical pressure working are no longer considered commensurate with the difficulties incurred. The boiler was taken up in 1929 by Messrs Siemens Schuckert G.m.b.H., Berlin, who persevered with the boiler and have installed quite a number of plants. The largest unit so far built was that in 1930 for Langerbrugge Power Station of Les Centrales Électriques des Flandres et du Brabant, Belgium. It was designed for critical pressure working, but it is now operated at pressures from 1662 to 2840 lb. per sq. in., super-heat 842° F. with an output of 222,000 to 275,000 lb. of steam per hour.

Several other boilers of the once-through-flow type have been brought out. We may mention, as having passed the experimental stage, the "Monotube" steam generator, 1933, of Sulzer Brothers, Winterthür, Switzerland, and the "Steamotive" boiler, 1935, the combined product of The General Electric Co., Schenectady, N.Y., The Babcock & Wilcox Co., Barberton, Ohio, and The Bailey Meter Co., Cleveland, Ohio.

The second type of forced-flow boiler is that in which there is continuous recirculation of 5 to 20 times the amount of water which is being evaporated during one circuit. This attractive idea had been, as we have seen (p. 162), worked out between 1859–61 by Martin Benson. The best known modern boiler is the La Mont, the invention of Commander W. D. La Mont, U.S.N. ret., who applied in 1918 for a patent for it which was granted in 1925 (No. 1,545,668). Owing to ill-success in the United States, the boiler was at a standstill in 1928, when it was seen by Dr A. Th. Herpen who believed that the principles it embodied were sound. He obtained from the assignees of the patent the world rights outside America and returned with it to Berlin, where he endeavoured to develop the boiler on a commercial scale. It is characterized by forced circulation through coils of small diameter tubing in parallel and to ensure that each coil has apportioned to it the appropriate amount of water, due to its situation, i.e. 8 to 10 times its evaporative capacity, distributing nozzles (see Fig. 77) are fitted in the inlets of the tubes. Each inlet is also fitted with a strainer to obviate clogging and so burning out of tubes. The pump is of the centrifugal type and to avoid more than one gland, which has of course to sustain the full boiler pressure, it is overhung. The pressure it has to maintain is equal to the resistance through the boiler circuit, usually about 35 lb. per sq. in. Apprehension as to pump failure has been met by the provision of pumps in duplicate. A diagram showing the general arrangement of a La Mont boiler is shown in Fig. 76. A boiler of 75,000 lb. of steam per hour, working pressure 815 lb.

per sq. in. and steam temperature 425° C., has been supplied to Imperial Chemical Industries, Northwich.[1]

The next main type to be considered is that employing indirect evaporation, i.e. the boiler water does not come into contact with the surface exposed to the radiant heat, but heat absorption from the furnace is effected by a liquid in a closed circuit which passes through coils in the water to be evaporated; having given

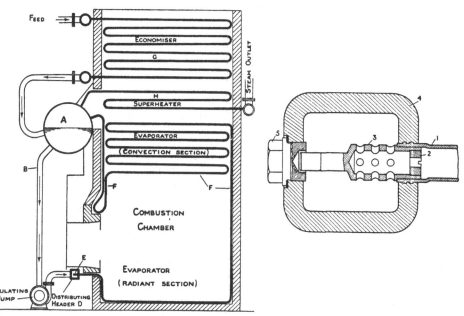

Fig. 76. Diagram showing arrangement of the La Mont steam generator circuit.

Fig. 77. La Mont distribution header and nozzle.

Courtesy of R. E. Trevithick, Esq.

off its latent plus sensible heat, the heat-carrying liquid is returned to the furnace by gravity or by a pump. A close analogy is that of the domestic hot-water heating system, where the radiators are not in circuit with the boiler but upon an independent circuit that gets its heat in passing through the hot-water tank. As regards the heat-carrying liquid, certain chemicals possessing valuable properties have been proposed but distilled water serves nearly as well and has obvious advantages. A foreshadowing of

[1] See *The Engineer*, March 25, 1938.

the idea underlying this boiler is the transmission of heat by superheated steam by Angier March Perkins in 1860, but in this case the steam is used for bread and biscuit making.

The best known of these indirect evaporation boilers is the invention of E. Schmidt and O. H. Hartmann, about 1930, known as the Schmidt-Hartmann and developed in Germany. At first sight it would seem that the closed circuit would have to be at a very high pressure in order to establish a sufficient temperature difference between it and the boiler circuit to effect satisfactory heat interchange. In a boiler erected in 1935 the pressures were 1420 lb. and 511 lb. per sq. in. respectively, but less difference is possible. A point of importance is that the closed circuit is composed of very long hairpin-bend coils of small diameter tubing exposed to the furnace. The circuit is extended into the evaporator drum and the heating element there may be of any suitable design, but it is of importance naturally that it shall be easily accessible; hence a manhole has to be provided in the drum. This has the disadvantage of making the scantling of the evaporator drum very heavy, and consequently it is of seamless forged steel.

Another indirect evaporation boiler is the Loeffler, invented and brought into a practical form by Dr Stephan Loeffler at the Vitkovice Iron and Steel Works, Czechoslovakia. It works on the principle that steam itself is forced by a centrifugal pump through the furnace-tube circuit, is superheated in the heat-absorbing area and gives out its heat in a direct-contact heat evaporator. A point of importance is that only steam of small specific volume is economical for forced-flow interchange; since specific volume decreases with rise in pressure, the higher the pressure the better. The pump has to stand the full pressure of the boiler, and the impeller is therefore overhung to save a second gland packing. The latter is of the labyrinth steam-packed type (cf. p. 223). Loeffler boilers have been installed in several European countries; one of the latest installations is that at the Brimsdown Station of the North Metropolitan Electricity Supply Co. The pressure is 2000 lb. and the steam temperature 940° F.

We now come to the rotating furnace boiler, the most sweeping and the most daring change in steam generation so far visualized. When water is in contact with a highly heated surface it assumes the spheroidal form and inhibits the evolution of steam. The assumption was that by making the boiler tubes to rotate at high speed the centrifugal action would break down the spheroidal state. The idea of making a boiler on these lines was mooted as early as 1851 by William Scott of Rising Sun, Indiana (U.S. Patent No. 8411 of 1851), and it has been carried into effect since 1924 in the "Atmos" boiler of J. V. Blomquist of Stockholm, Sweden. Here a squirrel cage containing the fire bars and the tubes rotates at about 330 r.p.m. Other boilers of the rotating type have been developed by H. Vorkauf, A. Th. Herpen and F. Hüttner, but they have not got far enough beyond the experimental stage to justify passing judgment upon them.

The pressure combustion boiler arose out of research carried out by Brown-Boveri & Co., of Baden, Switzerland, on supercharging of Diesel engines and the construction of the Holzwarth gas turbine. It was found that in nozzles for gas turbines, at velocities in excess of that of sound, a heat loss larger than was theoretically expected and increasing with the velocity took place. While detrimental to the performance of a gas turbine it was reasoned that heat transfer under such conditions might be used as the basis of a boiler. Such a boiler has been developed and the appropriate name "Velox" has been given to it. The stage in development reached at present can only be outlined (see diagram, Fig. 78). As might be expected only liquid or gaseous fuels have so far been found utilizable. Fuel by a special burner is injected into a circular combustion chamber at about 300 lb. per sq. in., while air is supplied at a pressure of about 35 lb. per sq. in., the latter being the pressure maintained in the chamber, making it possible to release 900,000 B.Th.U.'s per hour per cubic foot of chamber volume. To deal with this enormous heat release and velocity, the chamber is lined with evaporator elements through which water and steam are forced by a circulating pump; integral and concentric with the evaporator element is a superheater element. Thus the whole element receives direct heat on the outside and convection heat from the

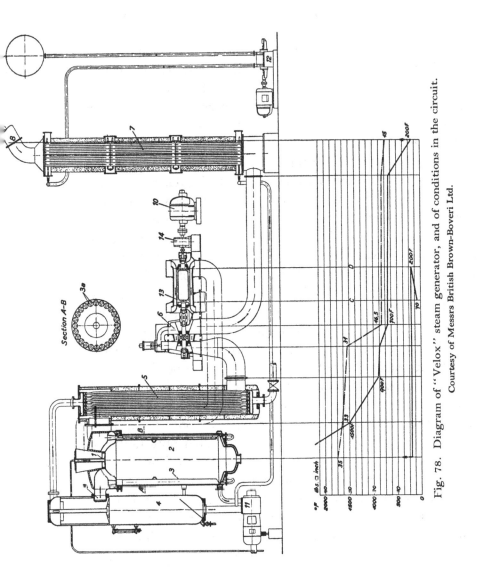

Fig. 78. Diagram of "Velox" steam generator, and of conditions in the circuit.
Courtesy of Messrs British Brown-Boveri Ltd.

gases traversing it inside. The flue gases pass through at about 900 ft. per sec. From the superheater the gases pass into a gas exhaust turbine which drives the air compressor. The gases then discharge into a contraflow economizer. The water in circulation is between 10–20 times the amount of steam generated. The steam-and-water mixture is introduced tangentially into a centrifugal separator which has a steam-releasing effect a hundred to two hundred times as great as the natural release in a horizontal drum, so that the resulting steam is dry. Further, automatic regulating gear is provided for (i) fuel supply, (ii) feed-water supply, (iii) auxiliary power supply. We are, therefore, at last within sight of an automatic steam-generating machine and not a mere boiler. The advantages claimed for the "Velox" boiler are high efficiency over a wide range, automatic control, and small cubic space occupied for a given output, making it applicable over a wide range of service.

It may be asked if it is not possible to dispense with furnaces and tubes altogether and heat the water directly. This has been done in the boiler invented by Oscar Brunler, father and son, in 1925 onwards. The fuel and air are burnt in direct contact with the water. Difficulties have arisen with ignition and starting, owing to the dirt imparted to the water, and the pressure of fuel and air supplies. The "steam" generated is a mixture of gases— steam, nitrogen, carbon dioxide and possibly hydrogen. How this mixed gas behaves and whether the difficulties mentioned have been overcome, the author is unable to say.

Whatever else may be said about the future of boiler development, the prospects are bright.

CONCLUSION

HERE we take leave of an absorbing theme after pursuing it through the centuries, and in doing so we would draw attention to the Synopsis of Events (facing p. 249), for on that chart is epitomized in chronological order the subject-matter of this volume, flanked by references to scientific discoveries and world history bearing on, or influenced by, the steam engine. To sum up. Newcomen in little longer than a decade developed the atmospheric engine; it fell into the hands of millwrights, fifty years passed almost without change and stagnation seemed to have set in. Then Watt came upon the scene and, in the space of a quarter of a century, not only improved the engine as a pumping apparatus but vastly extended its field of operations by making of it an engine to turn millwork. After that ensued a century of progress magnificent in its total achievement, but far from uniform in *tempo*; at first development seemed almost bewildering, for it took so many different directions; it rose to great heights and then slowed down, with an occasional spurt, to a steady jog-trot. Once more a brilliant period dawned with the advent of the steam turbines of Parsons and his coadjutors—this was half a century ago. Now, as far as we can judge, we are on the threshold of a period of consolidation and of marking time. The efficiency of a turbo-alternator is of the order of 85 per cent, and while obviously there is a long way yet to go thermodynamically, as the diagram, Fig. 54, shows, no form of heat engine offers more than slight prospect of being able to attain much greater efficiency.

Further progress is to be sought along some entirely new line of approach, as, for example, changing the potential energy of fuel directly into electrical energy. Powers of nature such as solar heat and the energy of the tides have been explored, but they require large outlay for plant, they are necessarily restricted as to area, and are intermittent in action, so that they could not become more than ancillary to, say, a grid system. The production of electrical energy from light by the selenium cell is still further in the background and suffers from the same drawbacks. The proposal to get heat from the interior of the earth, to details of which Sir Charles Parsons gave close attention in 1904, has hardly been pushed far enough for an opinion to be formed as to its practicability. As long as our fossil fuel lasts, these latter expedients will not engage concentrated attention, but when the fuel is gone they will come again to the front. Some persons may try even to concentrate lower grade heat into higher grade, but this is contrary to the second law of thermodynamics. We must not put aside another aspect of solar energy, viz. that made available through the agency of living organisms, whose efficiency in converting simple compounds like CO_2 and H_2O into complex organic compounds is astonishing.

So much for molecular energy which it has taken man many thousands of years to bring under control. There remains the hope, entertained for the last twenty years or so, that we may be able to release energy by nuclear disintegration—splitting of the atom as it is called. It is possible, and if we could get the energy in the form of heat, our present engines might serve; if in the form of radiations, some quite new engines would have to be developed. However, these appear to be idle speculations when we reflect that material and technical progress have outstripped our social and international consciences. The community would be well employed in making better use than it has done hitherto of the gifts that the scientist and the engineer have showered upon it; till then, further gifts would be in the nature of pearls before swine.

INDEX

Abraham, engine at Wheal, 100
Académie Royale des Sciences, 8, 50
Academies for the study of science, 3, 8
Ackton Hall Colliery, 227
Acta Eruditorum, 10–11
Adamson's flanged seam, 159–60
Ænigma, Prize, 51–3
Aeolipyles, 186–7
Air pump, Boyle's, 7, 9; Parsons's, 229; Von Guericke's, 6; Watt's, 69, 77, 92
Aktiebolaget, de Lavals Ångturbin, 211; Ljungströms Ångturbin, 219–20
Alban, high-pressure engine of, 98
Albion Mill engine, 83
Alexandria, seat of learning, 2
Alfred, engine at Wheal, 103
Allan, A., link motion of, 116
Allen, J. F., engine of, 142–3
Allgemeine Electricitäts Gesellschaft, 217
Architecture Hydraulique, 57
Atmos boiler, 243
Atmosphere, weight of, 6–7
Atmospheric engine, 30, 39, 54, 59, 61–3, 65–6, 90
Augmentor, Parsons's vacuum, 229
Austhorpe, engine at, 43, 51, 61, 62

Babcock & Wilcox's boiler, 168–9, 237–8, 240
Back, Mr, of Wolverhampton, 36
Balloon boiler, 116, 117
Barker's mill, 187, 190, 207
Barney, T., engraver, 37, 49
Barometer, invention of mercurial, 6
Barton, J., metallic packing of, 113
Bcake, T., 47
Beam of engine, 87
Bedplate of engine, 112
Beighton, H., F.R.S., 41, 43, 44, 52, 175
Bélidor, B. F. de, 57
Belliss's engine, 148–50
Benson's forced circulation boiler, 162, 240; critical pressure boiler, 239

Birmingham workmen assist Newcomen, 36
Black, Dr J., 68; doctrine of latent heat, 69, 175
Blading, turbine, 205–6, 218, 221–3, 230, 234
Blakey, W., engine of, 28; boiler of, 125–6
Blomquist's boiler, 243
Bloomfield Colliery engine, 73–4
Bodmer, J. G., stoker of, 132
Boiler, Atmos, 243; Babcock & Wilcox, 168–9, 237–8, 240; Benson's, 162, 239–40; Blakey's, 125–6; Blomquist's, 243; Brunler, 245; Church's, 131; Clarkson's, 128; Evans's, 93, 119; Eve's, 128; Field's, 170; Gurney's, 129; Hancock's, 130; Harrison's, 162–3; Herreshoff, 171; Howard, 165–6; La Mont, 240–1; Loeffler, 242; Monotube, 240; Newcomen's, 30, 35, 37, 40, 47; Niclausse, 170; Perkins's, 165; Root's, 167–8; Rowan's, 161; Rumsey's, 125; Savery's, 23, 25–7; Schmidt-Hartmann, 242; Serpollet's, 171; Steamotive, 240; Stevens's, 127; Summers and Ogle's, 131; Trevithick's, 92, 97–8, 119–21; Twibill's, 164; "Velox", 243–5; Wilcox, 168–9; Woolf's, 126–7
Boiler details, 159–60
Boiler efficiency, 172
Boilers, balloon, 117; Cornish, 120, 122; egg-ended, 120, 122; elephant, 127; forced flow, 239–41; haystack, 117, 122; indirect evaporation, 241–2; Lancashire, 121–2, 127, 159; locomotive, 123–4; pressure combustion, 243–5; rotating furnace, 242–3; waggon, 75, 118, 120; Yorkshire, 161
Boulton, M., F.R.S., 71; his Manufactory, 71; meets Watt, 71; buys Roebuck's share in Watt's patent, 72; partnership, 73; hustles Watt, 73; charges royalty, 74; insistent on rotative engine, 79; statement

CAMBRIDGE: PRINTED BY WALTER LEWIS, M.A., AT THE UNIVERSITY PRESS

Printed in the United States
By Bookmasters